Jan Apotheker
Teaching Chemistry

Also of interest

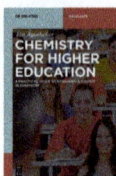

Chemistry for Higher Education
A Practical Guide to Designing a Course in Chemistry
Apotheker, 2018
ISBN 978-3-11-056957-5, e-ISBN 978-3-11-056958-2

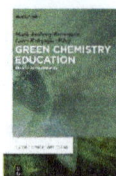

Green Chemistry Education
Recent Developments
Green Chemical Processing series, Vol. 4
Benvenuto, Kolopajlo (Eds.), 2018
ISBN 978-3-11-056578-2, e-ISBN 978-3-11-056649-9

Risk Management and Education
Meyer, Reniers, Cozzani, 2018
ISBN 978-3-11-034456-1, e-ISBN 978-3-11-034457-8

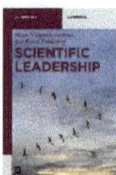

Scientific Leadership
Niemantsverdriet, Felderhof, 2017
ISBN 978-3-11-046888-5, e-ISBN 978-3-11-046889-2

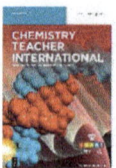

Chemistry Teacher International
Best Practices in Chemistry Education
Apotheker, Maciejowski (Editors-in-Chief)
e-ISSN 2569-3263

Jan Apotheker

Teaching Chemistry

A Course Book

DE GRUYTER

Author
Jan Apotheker
Retired from University of Groningen
Department of Chemistry Education
Nijenborgh 9
9747 AG Groningen
The Netherlands

ISBN 978-3-11-056961-2
e-ISBN (PDF) 978-3-11-056962-9
e-ISBN (EPUB) 978-3-11-056978-0

Library of Congress Control Number: 2018967156

Bibliographic information published by the Deutsche Nationalbibliothek
The Deutsche Nationalbibliothek lists this publication in the Deutsche Nationalbibliografie;
detailed bibliographic data are available on the Internet at http://dnb.dnb.de.

© 2019 Walter de Gruyter GmbH, Berlin/Boston
Typesetting: Integra Software Services Pvt. Ltd.
Printing and binding: CPI books GmbH, Leck
Cover image: Steve Debenport / E+ / Getty Images

www.degruyter.com

To Diederik and Rozemarijn

Preface

The aim behind the creation of this book is to make it a course book. Effort has been taken to make it a ready material that can be used either individually or by a group of students who wish to become chemistry teachers. I have therefore included assignments, which together will constitute a portfolio that can be used to present the qualifications and achievements of an individual student.

I have followed the development of a teacher, starting with a student giving his first lessons to a student becoming an experienced teacher. Chapters in the book can be used independently, as background material in a course. Some of the chapters can and will be used after a few years of experience.

https://doi.org/10.1515/9783110569629-201

Acknowledgments

This book is the result of about 20 years of experience as a teacher trainer and 25 years of experience as a teacher. Since then the approach to teaching has been changed gradually from being teacher centered to more student centered. This book is based on the courses I have taught and developed over the past 20 years together with many colleagues, both at my own university and in the Division of Chemistry Education of EuChemS and the Committee on Chemistry Education of International Union of Pure and Applied Chemistry. Without their input and discussion this book would not have been possible.

https://doi.org/10.1515/9783110569629-202

Contents

1 Introduction

Chemistry in secondary schools has gone through a number of changes in the past 20 years. Not only teaching and teaching methodologies have gone through changes, but also the content of chemistry education has changed. Things were not going well with the chemistry education. The appreciation for chemistry dwindled, the number of students electing chemistry at the university dropped to alarming numbers.

In "Science Education Now" (Rocard et al., 2007), an OECD Organisation for Economic Co-operation and Development report is quoted from which it becomes clear that the number of students electing a science and technology course dwindled by more than 40% in some cases between 1994 and 2003. In 2010 the Relevance of Science Education Report (ROSE) (Sjøberg & Schreiner, 2010) project reported about the relevance of science (and chemistry) education in schools for boys and girls.

It became clear that science education was not appreciated very much by large groups of students. Even though the attitude of students toward the roles and benefits of science appears to be positive. Some statements quoted from the ROSE project:
– *Science and Technology make our lives healthier, easier and more comfortable*
– *New technologies will make work more interesting*
– *The benefits of science are greater than the harmful effects that it could have*

These statements have been given to students who scored them around 3.7 on a Likert scale (Likert, 1978) of 1 through 5, indicating a high level of agreement. If you look at scores relating to science in school the results are completely different.

School science:
– *Is less interesting than other subjects*
– *Has not opened my eyes for new and exciting jobs*
– *Has not shown me the importance of Science and Technology for our way of living*

Low interest in subjects like:
– *Detergents, soaps and how they work*
– *Chemicals, their properties and how they react*

For boys only subjects like:
– *How petrol and diesel engines work*
– *Explosive chemicals*
– *The possibility of life outside earth*

seem to have some interest.

https://doi.org/10.1515/9783110569629-001

It is clear that the chemistry curriculum at that time was not very up to date. There was little or no link with the society. There were a number of main subjects that students were expected to learn and that had little or nothing to do with research at universities. The main subjects like equilibrium, acid–base theory and redox reactions, taught in secondary schools did not really clarify the position of chemistry as a molecular science, a science focusing on the study and control of molecular processes. Following other factors contributed as well (Yusuf, Taylor, & Damanhuri, 2017):

- *Teacher domination*
- *Curriculum content*
- *Perfectionism*
- *Competitive assessment*
- *Traditional teaching methods*
- *Poor environment*

These were reported as being factors with a negative impact on the quality of chemistry education.

At the same time the need for scientists and more specifically chemists, was made more explicit by public bodies like the UN and the European Union (EU). The EU formulated the Lund declaration (Union, 2015), which specified that Europe needed to work together to deal with a number of grand societal challenges like climate change, energy and water supply, public health and ageing societal changes in the world economy. This declaration has had major influence on programs like Horizon 2020. During the same time, the UN formulated its Sustainable Development Goals (Nations, 2016). These development goals are more concrete, than the goals formulated by the EU, and focus on scientific development that is needed to reach these goals.

In order to be able to work on these sustainable development goals, society needs to have enough scientists. Education in general and more specific chemistry education plays an important role in building these scientists (Pota, 2017).

This has set the scene for extensive work from several institutions to try and find ways to improve the situation of chemistry education; more specifically, trying to improve education in such a way that science becomes more attractive for students. Since early 2000, developments began on improving science education.

One of the first originated from the University of York, and was called "Twenty First Century Science" (Borley et al., 2016). Another was developed together with the American Chemical Society, and was called "Chemistry in the Community" (Powers, Langdon, Pentecoast, & Schwennsen, 2011). These texts had all the contexts from which students learned chemistry. The link with the "real" world was paramount in introducing new concepts. Another concept based on the idea of "need to know" was introduced as well. Only those concepts that are needed to understand a specific "real-life" situation are introduced. In "Chemistry in Context"

(Fahlman et al., 2017), a first-year college textbook, meant for students who were interested in chemistry, this was worked out even further, with subjects like "the air we breathe," "climate change" and "brewing and chewing."

Rocard et al. (2007) published a report on behalf of the EU about the situation in science education. They came up with a number of suggestions.

One of them was a switch in science teaching pedagogy from more traditional methods to inquiry-based methods. Also important is that teachers are considered to be key players in the renewal of science education. The development of networks for professional development deemed an important form of teacher training.

A year later, Osborne and Dillon (2008) also reported a number of recommendations on science education. They emphasized on the relation between science education and the material world, more than being a foundational education for future engineers and scientists. Innovative curricula that focus on low student motivation should be stimulated. The use of contexts in understanding science are mentioned specifically. In this report, the role of science teachers as major stakeholders in the development of high-quality science education is emphasized.

This has led to several programs within the Framework Programs of the EU (Reillon, 2017), notably in the FP7 program (Anonymous, 2012) as well as within Horizon 2020 (Commission, 2017). In these programs, the introduction and development of inquiry-based science education was an important factor.

These programs have stimulated the development of alternative programs. In a special issue of the *International Journal of Science Education* (Gilbert, 2006), context-oriented science education was emphasized, as a way to improve the perspective of students toward science.

In several countries, developments resulted in a new context-oriented chemistry curriculum (Apotheker, 2008; Nentwig, Demuth, Parchmann, Gräsel, & Ralle, 2007).

Some of these developments have led to new curricula (Apotheker et al., 2010) or the formulation of new standards (Talanquer, 2014).

Another important for the changed science education was the publication of the book "How people learn" (Bransford, Brown, & Cocking, 2000). This book summarizes the findings from research about learning. This has influenced ideas about the relationship between teachers and students. It has brought about quite a few innovations in the practice of teaching.

As a prospective teacher you will be in chemistry classroom about 10 years later after publication of this report and will experience some of these changes. For your development as a teacher the background of these changes, as described previously, are important. They also indicate the importance of keeping up with the literature about the development of chemistry education.

2 First steps to become a chemistry teacher

2.1 Required chemistry background

You are about to take the first steps for a job as a chemistry teacher. Before choosing this as your career, you must have learnt chemistry. You might have taken a bachelor's or master's degree in chemistry or chemistry education. Either one is fine. You need a certain level of active knowledge in chemistry before you start teaching chemistry in upper secondary school. For lower secondary school the requirements are less.

The active concept knowledge you need can be found in textbooks that are generally utilized in college for introductory chemistry courses. Examples are

"Chemistry" (Blackman, Bottle, Schmid, Mocerino and Wiile, 2011) or "Chemistry," the molecular nature of matter and change (Silberberg & Amateis, 2015).

This is also the minimum knowledge required for teaching chemistry in lower secondary school.

For higher secondary education, you should also have a strong background in the following:
- Physical chemistry
- Quantum mechanics
- Thermodynamics
- Inorganic chemistry
- Organic chemistry
- Biochemistry
- Macromolecular chemistry
- Green chemistry
- Materials

Apart from the basic knowledge, you should have studied one subject deeply and done a research project for at least six months.

This means you need to have a chemistry or chemical technology major, although pharmacy majors, environmental majors, molecular biology majors can also qualify. Even though in some related majors, a few extra subjects might have to be studied. In most cases, these are some related majors that need to be taken as an undergraduate course:
- Quantum chemistry
- Macromolecular chemistry
- Green chemistry

https://doi.org/10.1515/9783110569629-002

Assignment 2.1

Basically in upper secondary school, you should be able to understand and explain the problems faced in the International Chemistry Olympiad (IChO) (Nick & Nather, 2007). As a teacher in higher secondary education, you will be expected to guide your students to participate in that competition. You can find these problems through the website of the IChO (Sirota, 2015). This site will also contain a link to the website of the current Olympiad. Each year the national organizing committee publishes a set of preparatory problems that are used for the preparation of the Olympiad competition. If you download these set of problems for the current or the last year, you will get an idea of the level. Apart from this, you can use these problems as a source for assessment in the classroom. Generally, they are well thought-out problems. A scientific committee consisting of leading research chemists in the organizing country normally takes at least a year to make them.

Assignment 2.2

After conducting research for a year or two in a specialized field, your background knowledge on the concepts of chemistry taught in secondary school will not be activated. By studying one of the textbooks and solving some of the IChO problems, this can be solved easily. The level of the textbooks mentioned is such that with the theory discussed there, you should be able to solve most of the IChO problems.

2.2 Beginning as a teacher

When you begin as a teacher without any prior training, it will be like jumping into deep water and not knowing how to swim. Or it will be like the first time you drove a car. You have no idea what to do and how to do it. If possible try to take an introductory course in teaching before you begin your formal apprenticeship as a teacher. Learn some basic facts first. In an introductory course, you should be admitted to a secondary school and have a teacher as the coach. Working as a teaching assistant in university helps but facing a group of say 25 12-year-olds is a completely different matter.

This chapter should help you overcome some of the problems you will experience while beginning as a teacher. It will give you a chance to begin in a responsible manner. Compared to learning how to drive, you will have someone next to you as a coach, who is also at the controls and can take action when something goes wrong.

Before you start teaching yourself, you should have at least observed a number of lessons. Not only lessons are taught by your coach, but also lessons are taught by colleagues of chemistry, physics and biology. You will then get an idea of what they do and how they do it.

Each teacher has his or her own way and style of interacting with a class. You will have to develop your own style. This is why it is important to observe other colleagues. You should include observing your mentor or coach for the whole day. Following your mentor or coach for one full day, will also give you some indication of the workload of a teacher.

Assignment 2.3

In order to find out what it means to be a student in secondary education, it is a good idea to follow the class you will be interacting with a teacher as an apprentice, during one full day of lessons. This will give you an idea of their behavior as a class. This will help you when you begin your first activities. It also gives you an idea of how tiring a day in school can be.

Assignment 2.4

After these observations and these assignments, you should have a basic idea of a number of aspects involving teaching, both from a teachers' standpoint as well as that of the students.

2.3 Roles of a teacher

When you stand in front of a classroom and have the responsibility for a group of students during a lesson, you take up several roles. The number of roles described will vary somewhat, but here five different roles are discussed (Slooter, 2009). These roles will help you focus on what is expected of you as a teacher. Again, if compared with driving, you need to steer, are responsible for safety, have to use brake, clutch and gear and so on.

In teaching, your roles include the following:
- *Host*
- *Presenter*
- *Didactic expert*
- *Pedagogue*
- *Closer*

These roles are loosely based on five dimensions of learning formulated by Marzano (Marzano & Association for Supervision and Curriculum Development VA., 1992).

These dimensions are as follows:
- *Developing positive attitudes and perceptions about learning*
- *Acquiring and integrating knowledge*
- *Extending and refining knowledge*
- *Using knowledge meaningfully*
- *Developing productive habits of mind*

The role of host implies that you welcome your students in the classroom, put them at ease and give them the idea they are in a safe environment. You should be present in the classroom before the lesson starts, standing near the doorway, making eye contact with the students as they enter, radiating self-confidence. Welcome the students by their name, follow the same ritual every time. Be well prepared before starting a lesson. When everyone is in and seated make sure they sit according to your rules:
- No satchels on the table
- Books and relevant papers on the table
- No smartphones in sight
- No outdoor clothing
- No caps or other head covering (although for females you may have religious problems here)

All these rules indicate that you welcome your students, you respect them, but you expect them to respect your criteria. When you observe your colleagues, you will find they perform this role in different ways. You will need to find a way that fits your own personality. However, the start of this lesson is extremely important for the rest of the interactions between you and your students.

The second role is that of a presenter or perhaps the chair is a better description. You are responsible for all the happenings in the classroom. You are basically the leader. The problem is that this will not happen just like that. Unlike an audience where this goes automatically, students in a classroom need to be convinced that you are indeed the one that is chairing this session. This means you need to make clear to the students that you are taking on that role. This means you have to radiate that you are taking charge, by saying something, or clicking your fingers, knocking on the table and whatever to draw their attention.

Saying something like: "I would like to start the lesson, so please be quiet" works fine, if you at the same time demonstrate that you fully expect the students to comply. If things do not work out as you expect them to, you will need to revert to the role of the pedagogue.

The third role is that of being the didactic expert. This means that you are the expert in your subject. You know more than enough to be able to let the students learn; not only about the subject but also about the most effective ways of learning. You are the person to guide the learning processes of the students. You do that in several ways, you can give a short presentation, using the blackboard or PowerPoint to support the presentation. As a chemistry teacher you may given a demonstration.

There are many didactic tools you can use which will be discussed at later. Again, here if you want the students to be active then make it clear that you expect them to act. Depending on the didactic tool you use you will need some sort of action to get the students to work.

The fourth role is that of a pedagogue. In this setting, it means that you are able to manage the classroom in an effective manner. Are you able to deal with the students in such a way that they display the correct behavior? Most important question: Are you able to correct the deviant behavior of your students? For a beginner as a teacher, this is one of the most challenging roles to fulfill. It is very hard to define effective behavior that will correct students. What you need to do depends very much on the age group as well as the level of the students. The most important part is that you show clearly that in your behavior you make clear that you are determining what happens in the classroom. In the insert, a short description is given of what the role of attitude can be, related in a story and told in a group of the author's students.

In an introductory course on teaching chemistry, we always start with an "peer-to-peer coaching" session. In this session, students are asked to relate their experiences in the classroom. One student was complaining that there was a group of three students who were constantly disturbing the lesson. He or she could not get a grip on them. In the discussion that entailed, possible disciplinary measures were discussed. The ultimate one is expulsion from the lesson was something he or she had not yet used. In the end, it was decided he or she would expulse the three students after one warning. He or she would report a week later what had happened.

A week later, he or she came back and explained: "I was completely surprised. It was as if they knew I was going to expulse them if they did not listen, so they behaved themselves."

The attitude of the students in front of the class played a major role in the above-mentioned example. Something that may help understand the problem was another remark that one of the students made at another session:

...you are not going to be intimidated by a bunch of adolescents are you?...

The role of pedagogue is one of the roles that is often one of the most difficult ones to learn. You will need to find out what works for you. One of the main problems especially when beginning as a teacher is the part from all your other roles that you need to be able to keep an overview of what around 25 kids are doing. Steering that behavior is one of the challenges as a beginner teacher. For effective teaching and for creating a safe and secure learning environment, it is absolutely vital that you are able to manage the behavior of the class full of exuberant adolescents.

The final role to play is that of the closer. This is actually part of being a presenter or host, but it is a different aspect. You opened the lecture session, you will need time to close it. For the learning process, it is important that you summarize what was done during the lessons, what the learning goals were and what the students need to do for homework. You have to realize that the time you have contact with your students is only about a third to one-half of the time they are expected to spend on your subject. So, homework as well as preparing for the next lesson is important. Given the age group of the students, you will need to make sure they note exactly what you expect them to do for the next lesson.

2.3.1 Feedback

Getting feedback from a coach or a fellow student about the way you fill into these roles is extremely important or learning to effectively take up these roles. Make sure you set yourself learning goals at the beginning of each lesson. Ask observers to specifically comment on these learning goals. If you want them to observe a role, ask them to do so. Interact with the students to find out how they have experienced your activities. Both your own reflections as well as those of your coaches and peers will help you learn to become more effective in your role of a teacher.

There are many forms available for receiving feedback, depending on the type of feedback you would like to have.

2.4 First teaching activities

Normally, you will start with giving perhaps a 10-min instruction during a lesson. This could be the introduction of a chemistry concept or the introduction of a lab session. It will give you the opportunity to focus on one particular role as a teacher, in case the one of a presenter and the didactic expert. Because the regular teacher is present, you need not worry about the other roles. Of course, you need to prepare those 10 min, and evaluate them afterward with the teacher. Forms are very often used to prepare yourself. Table 2.1 is an example of such a form.

Apart from some data, you need to indicate what you are planning to do and what you expect your students to do. This can be: "*being silent, listening and paying attention.*"

An assessment is needed to find out whether or not the students understood what you wanted to teach them. This can be done by letting them solve a problem or responding to questions. Assessment will be discussed in a chapter.

Normally, you will discuss this plan with your coaching teacher. He or she will advise you about your plans.

Receiving feedback from the teacher is important. He or she will be using a form as well to take notes. An example is given in Table 2.2.

These forms are a help to structure the discussion with your coach or teacher.

In the discussion with the teacher who is coaching you there will be a number of standard questions:

Table 2.1: Preparation form for a short introduction.

Name school:	Class:	Teacher:
Apprentice teacher:		
Sections in the textbook:		
Concepts to be discussed:		
Assessment to be used:		
Classroom activities:		

Time:	Teacher activity	Student activity	Notes (How did this go?)
0–5 min			
5–10 min			
10–15 min			

- How did you think the instruction was received?
- Which parts are you happy about?
- Which parts could be improved?

After that he or she will comment on the things he or she observed. This will help you improve on your interaction with students.

Assignment 2.5

After you have given a few introductions, you will move on to teach full lessons. Your preparation should be more complete. During a full lesson, you will need to carry out several learning activities. In Chapter 4 the design of a lesson and series of lessons will be discussed in more detail. This will include the discussion of several types of activities. You will need to include learning goals for a full lesson. Learning goals are basically a description of what you want the students to learn. You will need to assess whether or not the students have achieved these learning goals. The form from Table 2.1. will be extended somewhat in order to do that. This form is described in Table 2.3.

You will need to take into account that the attention span of students is about 5–10 min. This means you will need to switch to another activity after a maximum of 10 min. This entails that you switch between teacher roles. That is why these are included in this form.

A more detailed observation form that could be used is given in Table 2.4.

Table 2.2: Evaluation form for a short presentation.

Name school:	Class:	Teacher:
Apprentice teacher: Use the following categories for evaluation: 1. Predominantly weak – 2. More weak than strong – 3. More strong than weak – 4. Predominantly strong – 5. Not observed –		
Was the student well prepared?		
Did the presentation go as planned?		
What went well?		
What could be improved?		
Specific items:		
Nonverbal communication		
Eye contact with students		
Positive attitude		
Radiating self-confidence		
Relation with students		
Verbal communication		
Voice used well		
Interactive with students		
Level of language used		
Teacher roles		
Host		
Presenter		
Didactic expert		
Pedagogue		
Closer		

Table 2.3: Preparation form for the lesson plan of a whole lesson.

Name school:	Class:	Teacher:

Apprentice teacher:

Sections in the textbook:

Starting knowledge of the students:

Importance of the lesson for the students

Learning goals to be achieved:

Core activity of the lesson:

Media to be used:

Assessment to be used:

Personal learning goals for this lesson:

Classroom activities:

Time:	Teacher activity	Student activity	Teacher role	Notes (How did this go?)
0–5 min				
5–10 min				
10–15 min				
15–20 min				
20–25 min				
25–30 min				
30–35 min				
35–40 min				
40–45 min				

Table 2.4: Detailed observation form for a whole lesson.

Name school:	Class:	Teacher:
Apprentice teacher:		
Use the following categories for evaluation: 1. Predominantly weak – 2. More weak than strong – 3. More strong than weak – 4. Predominantly strong – 5. Not observed –		
Was the student well prepared?		
Did the presentation go as planned?		
What went well?		
What could be improved?		
Specific items:		
Communication aspects		
Is passionate		
Uses voice and body language effectively		
Is friendly to students		
Pedagogical aspects		
Creates relaxed atmosphere		
Supports self-confidence of students		
Gives students attention		
Corrects unwanted behavior		
Supports wished behavior		
Stimulates students		
Didactical expertise		
Displays professional knowledge		
Explains relevance of activities		
Is able to maintain logical structure		
Is able to instruct students clearly for each activity		
Is clear in his instruction		

Table 2.4 (continued)

Name school:	Class:	Teacher:
Is able to answer questions clearly		
Is able to coach students during learning activities		
Checks learning progress of students		
Presenter		
Is clear about the startup of the lesson		
Manages time well		
Is well organized		
Activates students		
Involves students		
Uses media effectively		
Closer		
Summarizes learning goals		
Summarizes what happened during the lesson		
Is clear about work to be done before the next lesson		
Evaluates lesson with students		
Student behavior		
Students are respectful to teacher		
Are actively engaged while listening		
Are actively engaged with other lesson activities		
Self-analysis		
Is open for feedback		
Asks feedback on specific issues		
Is able to analyze strong and weak points		
Is able to formulate improvements		

Assignment 2.6

To evaluate a lesson or part of a lesson Table 2.5 can be used.

Table 2.5: Evaluation form for (part of) a lesson.

Name school:	Class:	Teacher:
Apprentice teacher:		
Lesson observed by:		
Date of observation:		
Deviations of lesson plan		
Things that went well		
Things that could be improved		
Intentions for next lesson		

You should try to finish an introductory course with teaching preferably four to six consecutive lessons. During the first part, your teaching coach should be present during the lessons. You will need to discuss your preparation with him/her. For each lesson, you will need to prepare a lesson plan. In the evaluation of each lesson, try to get feedback about the way you fulfill each role of a teacher. Try to teach the last two lessons while your mentor orcoach is not present during the lesson. Your teacher role as a pedagogue will be more challenging when the regular teacher is not present.

Assignment 2.7

2.5 Positions in the school during an internship

2.5.1 Apprentice teacher

As an apprentice teacher, as described in the previous section, your position is directly linked to your mentor or coach. He or she is responsible for you and your actions. You have no formal position in the school as such. Your coach or mentor should introduce you to director of the school, and other teachers and teaching assistants. If possible, your coach should let you accompany him/her to meetings like a report meeting about the results of students in a class, or a meeting of the chemistry group. But again, your position is not an official one within the school structure. An apprenticeship as described in the previous section normally takes about three to six weeks. You will observe in total about 10–20 h of lessons. You will be teaching about 20 (parts of) lessons. The total time involved in this stage will include discussions with your coach teacher before and after each lesson. It should add up to about 140 h of study time. In a university, the apprenticeship will run parallel to an introductory course of pedagogy and teaching methodology. This theoretical course will also involve about 140 h of study time.

The apprenticeship is intended to prepare you for the following step in becoming a teacher.

The learning goals you should achieve include fulfilling the roles of a teacher. Apart from this, you should be able to function in the team of colleagues working in the same discipline, as well as with other colleagues in the school. Most important, you should be able to reflect about your own activities in the classroom, and your development as a professional teacher.

2.5.2 In-service teacher

This step involves an in-service training arrangement, in which you function as a teacher and have responsibility for two or more classes. You will normally receive training at the university as well as training in school. This type of training normally takes at least a year and includes several theoretical courses.

2.5.3 Mentor/coach

One of the chemistry teachers will be assigned as your mentor orcoach. He or she will introduce you to most of the tasks you need to perform. He or she will also help you prepare for your lessons, assessment and so on. He or she will attend your lessons every so often and give you feedback. You can also ask him or her to attend lessons and ask him or her specific feedback on a lesson. He or she is your prime contact in the school.

2.5.4 Colleagues

Teachers are organized in departments in a school. You will be part of the chemistry department. That is most likely linked to the department of other sciences. Your mentor or coach will also be part of that group. Within the department, the year programs and the overall schedule of subjects to be covered during the year are discussed. Any coordinated assessment test for a whole year group will be discussed as well. The department will have a set of assessment test that have been used in previous years.

You will be part of the team that teaches chemistry and will cooperate with other teachers in preparing and teaching students in different year groups.

2.5.5 Teaching assistant

Not all countries will provide a technical assistant for teaching science. But it is a great help if there is one. He or she will be part of the department. He or she will be experienced with all lab assignments that are used in the school. He or she is the person who coordinates all lab activities. You will need to arrange with him or her any lab you wish to do with your class. He or she will prepare everything and also be in the lab to assist you and the students. If you want to try out anything new, you will have to discuss that with him or her beforehand.

2.5.6 Rest of the school

You are of course now part of the whole school organization. So, you will meet the director, the other administrators and your colleagues of other departments. Every so often there will be a teacher meeting which you need to attend. There will be departments as well as inter departmental meetings.

2.5.7 Parents

The parents have entrusted their children in your care. You are personally liable for their care. The school will have some type of insurance, as well as the institute that is training you. But make sure you have some sort of personal liability insurance that covers this.

Once or twice a year parents will have the opportunity to make an appointment with you to discuss the progress of their child.

2.5.8 Extracurricular activities

All schools organize extracurricular activities. These range from having an orchestra, organizing a play to organizing excursions.

It is a good idea to try to get involved in one of these extracurricular activities. You will get to know the other faculty members better, as well as the students. Your participation will be greatly appreciated in general.

Assignment 2.8

Assignment 2.1 Chemistry biography

Write a 500–1,000-word paper describing the chemistry you studied and why you find chemistry such an alluring subject. Indicate what fascinates you in chemistry and explain why you wish to become a teacher.

Hand in the paper to your instructor and to at least two peers. Ask them for feedback.

Describe in 200 words your reaction to their feedback.

Include the feedback as well as your reaction in a portfolio.

Assignment 2.2 Preparatory problems of the IChO

Download the set of preparatory problems of the latest IChO (they are normally published in the second week of January).

Analyze the content knowledge needed for each question and indicate whether or not you would need to study the background knowledge in order to answer the question. Discuss and compare your analysis with your peers.

Work out at least two problems of the IChO.

Add the result of this assignment to your portfolio.

Assignment 2.3 Observing your coach and other teachers

When you observe your coach or mentor, you should pay attention to a number of things.

His or her interactions with the class, relating to the way he or she carries out the different roles of a teacher are important.

For his or her role as a host the first 5 min are important, because they set the stage for the rest of the lesson. When you observe some of the other science colleagues or you are in classes of other teachers pay attention to this as well.

For the role as a didactic expert and presenter, observe how your coach is instructing the class. Note any questions and discuss these with him or her afterward, specifically trying to find out his or her motivation for following a certain line of instruction.

For the role as a pedagogue, observe how he or she interacts with students and reacts to situations in which students behave in a way he or she does not like.

Finally, observe the way he or she closes the lesson and how much time he or she takes to do that.

Write a short (300 words) report about your observations and put it in your portfolio.

Discuss these observations with your fellow students.

Assignment 2.4 Observing a class for a full day

You will be interacting with one specific class during an introductory course. This is the class you should follow during the full day. Observe the behavior of the class in different lessons. Before you attend a lesson, you will have to ask permission of the teacher teaching this class. In order to facilitate thia process, ask your mentor or coach to inform the teachers of that class beforehand.

You are bound to observe conflict between a teacher and one or more students. The teacher will resolve this conflict in a certain way. Afterward, interview the teacher about the conflict and find out why he or she intervened the way he or she did. Do the same with one of the students involved in the conflict. Check how the students rate the way the teacher intervened.

At the end of the day, interview the student mentioned previously, as well as a quieter student. Find out how they experienced the day. Ask what they liked and did not like and find out why.

Write a short report about your observations (500 words) and discuss the report with at least two peers and your instructor. Put the report and the received feedback in your portfolio.

Assignment 2.5 Preparing a 10-min instruction

Prepare a lesson plan for your instruction, using Table 2.1.

Discuss the lesson plan with at least two peers and your instructor. Ask your coach to use Table 2.2 for the observation and feedback.

Fill in Table 2.5 for evaluation.

Add the forms to your portfolio.

Assignment 2.6 Prepare a single lesson

Use Table 2.3 to prepare a whole lesson. Ask the observer to use Table 2.4. as an observation form.

Evaluate the lesson with your observer and fill in the form of Table 2.5.

Add this to your portfolio.

Assignment 2.7 Preparing a lesson series of 4–6 lessons

Prepare a lesson plan for six lessons indicating at least:
- The sections in the textbook you will be covering
- Indicate the problems or exercises you want the students to make
- Indicate the lab work you want the students to do (if any)
- Indicate the knowledge the students should have before the start of the lesson series
- Indicate the learning goals you wish to achieve in these six lessons
- Indicate how you will assess whether you reached the learning goals
- Make a lesson plan for each lesson you want to teach

Reflect on each lesson and indicate what you will change in the next lesson as a consequence.

At the end, look back on the lesson series and indicate what you will change next time.

Discuss this whole report with your coach or mentor as well as at least two peers and your instructor. Use Table 2.5 for the evaluation of each lesson.

Include all forms into your portfolio.

Assignment 2.8 School organization

Prepare a short report or powerpoint about the organization in your school to share with your peers. During the lecture session, compare the organization in your school with that of your peers.

Add the report or powerpoint to your portfolio.

3 Introducing chemistry in lower secondary schools

3.1 What is science?

As a chemistry teacher, your role as a didactic expert is to introduce chemistry to students.

You are the person providing chemistry content to students. You do this by interacting with the students about chemistry. Introducing them to new concepts and ideas, most students never thought about or realized that this existed. In order to do that, you need to introduce the concepts of science or more precise natural sciences. This relationship is depicted in Figure 3.1, called the didactic triangle. It relates the role you have as a teacher as an intermediate between the content and the student. You determine at what time specific content is introduced to a student, as well as the way in which this is done. These decisions are based on your didactic knowledge of both content and the learning process of the student. Hence the title 'didactic triangle'.

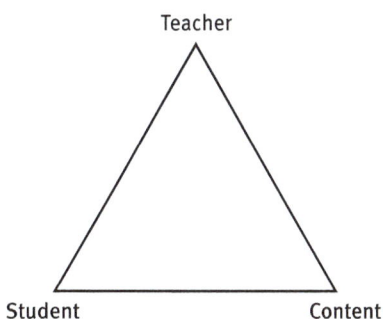

Figure 3.1: Didactic triangle. (From Kansanen & Merk (1999).

The nature of science is one of the first subjects that will be discussed when starting to learn about chemistry. Nature of science is a broad concept (Laherto et al., 2018). Over time, this concept has changed considerably. It is a broader combination of historical, philosophical and sociological ideas about science. It is not the same as the scientific method, which is based on the nature of science. The nature of science not only encompasses chemistry, but all other sciences as well. It also discusses the role of science in society.

One of the main factors in the nature of science is that science is an experimental science. Scientific knowledge is based on experiments and the interpretation of experiments. Because of this, scientific knowledge is considered to be tentative.

The nature of science is something that keeps on evolving. The primary focus of science has been the explanation of natural phenomena. In classical times, Plato

https://doi.org/10.1515/9783110569629-003

and Aristotle are examples of thinkers who shaped scientific ideas, using logic as a major tool for finding explanations for natural phenomena. It was not until Galileo Galilei published his dialog (Galilei, 2001) that experiments were considered more valuable than logic. Experiments took place of course in ancient times, but the "truth" was found through logic.

Over the years, science changed. Especially in the nineteenth century, philosophers discussed science as a means to arrive at "truth" (Shaw, 2013). Experiments played a vital role in arriving at the truth.

Since Lavoisier (late eighteenth century), chemistry became more and more an independent science. The nineteenth century was the time in which chemistry separation techniques were refined, which led to the discovery of many elements and ultimately to the publication of the periodic table in 1849 by Mendeleev. Skills in organic synthesis were developed at the same time; for example, Liebig played an important role. With the increasing technology in the nineteenth century, experiments gave more and more insight into the processes at the molecular and atomic scale. This led to a different insight into chemical processes. In order to bring some unity in the language and symbols used in chemistry, IUPAC was founded in 1919 in Paris.

In the early twentieth century, regular surface mail and face-to-face meetings were the only way by which scientists can communicate. That is why the Solvay Conferences, which started in 1911, were so important. The world's leading scientists, including 17 Nobel prize laureates in 1927, were present at these conferences to discuss the scientific problems faced that time (Figure 3.2).

When CERN Conseil Européen pour la Recherche Nucléaire in Geneva opened up its communication channel for the world, Internet was introduced. With the introduction of "Internet" the exchange of information took an enormous leap forward. The current development of open access makes the sharing of experimental results a lot easier.

In the twentieth-century philosophical discussion about the role of science and scientific inquiry continued (Shaw, 2013). In particular, Karl Popper introduced the idea that experiments should be focusing on refuting a particular hypothesis. Science in that sense has primarily a task to eliminate false theories. Popper coined the phrase pseudoscience, meaning a field in which this type of experimentation did not take place. An example of such a field is homeopathy.

Kuhn (Domin, 2009) introduced the term paradigm. He indicated that paradigms are the broad conceptual and methodological orientations shared within a science. Research questions are mostly determined from the background they are set.

This last theory is what has influenced most teachings of the scientific method, which has the following steps:
- Formulate a scientific question
- Formulate a hypothetical answer to the question

Figure 3.2: First Solvay Conference.

- Design an experiment that can yield information to answer the question
- Carry out the experiment
- Analyze the data
- Discuss whether or not the data provide an answer to the questions
- Conclude what other experiments may be needed to answer the question

Normally, these questions will be imbedded within the existing knowledge and will seek answers to details that are not clear yet.

Normally, the nature of science will be part of the curriculum in lower secondary education. It will continue when students elect a science major in upper secondary. Within discussions about science, models are often used to clarify the discussion. The use of models is something that should be explained carefully in lower secondary school. It should be made clear that models are incomplete representations of reality that are used to explain and predict observed phenomena. Most important factor is that models can be adapted when new facts arise. This implies that models are not constant. It also implies models are not true or false. Models are more or less adequate representations of reality. When discussing models like the atomic models in chemistry, it should be made clear that the Bohr model of atom is a refinement of Rutherford's model, which is a refinement of Dalton's model.

3.2 What is chemistry

Chemistry is introduced at different levels in education. Most often somewhere in lower secondary, students will learn about chemistry for the first time. Very often they will have had some instruction in biology and physics.

In biology, nature will be taught, describing and scaffolding both flora and fauna. Ecosystems are sometimes discussed in biology. Very often the food chain will be discussed. Biology at this stage is taught at the organism level. Students also learn about their own bodies, learn about hygiene, sexual reproduction, illnesses and so on.

In physics, the main subjects are describing and bringing order in everyday events, like movement, acceleration, distance covered, heat, calories, light and lenses.

Chemistry is normally introduced after the introduction of physics and biology.

The first questions asked in chemistry education are normally about the difference between biology, physics and chemistry. An answer to this question is not always easy to give, as the differences are sometimes subtle. In the past the difference was related to chemical reactions. In chemistry, you study what happens when substances react with each other and yield new substances. In physics, substances do not change. In biology, you study organisms and the relation between organisms.

This idea has changed somewhat with the introduction of Johnstone's triangle (Figure 3.3).

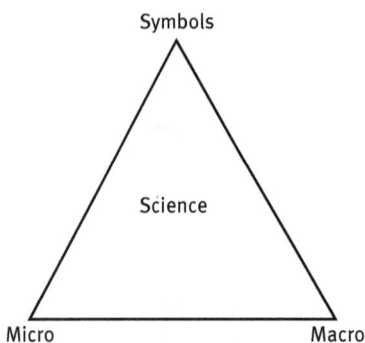

Figure 3.3: Johnstone's triangle (from Johnstone, 1997), based on the triangle in the article.

This triangle describes the relationship between different levels of size when studying science.

On the right, macro signifies phenomena that occur at the macroscopic scale, meaning phenomena that can be observed directly. The term micro is slightly confusing as it indicates processes and phenomena at the molecular and atomic level. Symbols are symbols and formulas that can be used to describe processes and relationships that are occurring at the macroscopic level or at the microscopic level (Johnstone, 1997).

In terms of size, micro indicates processes at a scale of about 1 nanometer and smaller. Macro indicates processes from about 1 micrometer and larger. Basically, processes that can be seen directly are considered to be macroscopic phenomena. Since 1997, nanotechnology has come up as a separate science, which has led to the definition of a mesoscale (Figure 3.4).

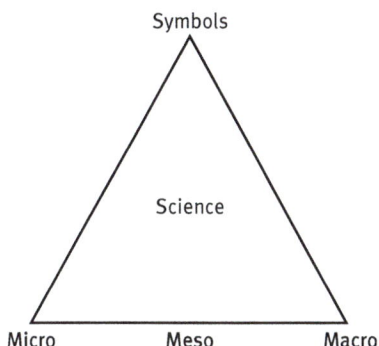

Figure 3.4: Johnstone's triangle with mesoscale.

This mesoscale indicates processes that occur at an intermediate level, of approximately 10 to 100 nanometers, which is the scale that is studied in nanotechnology. It has become clear by now that the properties of substances at this scale differ from the properties at the micro- or macroscale.

This triangle can be used to define the separate sciences, but also shows the overlap of the sciences. In terms of this triangle, mathematics is the science that studies symbols and formulas and the way these can be manipulated in a logical manner.

In physics the macroscale is studied. The symbols and formulas are used to describe the macroscopic phenomena. These symbols and formulas are used to predict what will happen in certain situations.

Biology studies relationships between organisms at the macroscopic level. Formulas and symbols are not used overly much in biology. The scale is at the level of organisms. This changes when processes are studied at the cellular level. At a certain point, it becomes molecular biology and overlaps with chemistry. For physics the same applies as soon as subatomic processes are studied, for example.

In chemistry, macroscopical phenomena are linked to processes and properties at the micro level, that is, the molecular and atomic level. In chemistry, symbols are used to describe processes either at the macroscopical level or at the atomic/molecular level. This definition is broader than the definition of chemistry given previously, involving chemical reactions (Figure 3.5).

This definition, in which chemistry studies the relationship between processes at the macroscopic level and linking those to processes at the atomic/molecular level, should be at the core in teaching chemistry in secondary education. This means that it must be made explicit to the students when the macroscopic level is introduced and

Mathematics
Symbols

Physics

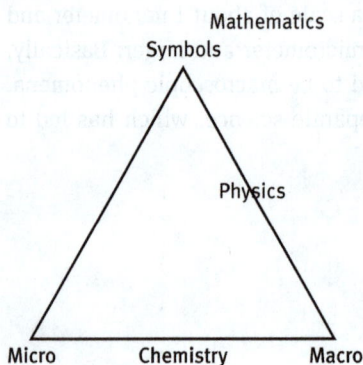

Micro   Chemistry   Macro   Figure 3.5: Subjects related to Johnstone's triangle.

when the related atomic/molecular level is discussed. When comparing the surface tensions of 1-butanol and 1-ethoxyethane a macroscopic property is meant, which can be explained using the polarity of the molecules involved at the microscale.

Equally important is using the correct symbols for describing the state you are discussing. Sometimes there are subtle differences. $N_2(g)$ denotes nitrogen at the macroscopic level. N normally means an atom of nitrogen, while with N_2 a single nitrogen molecule is meant, relating to the microlevel.

In most textbooks, this distinction between macro and micro is not made very explicit, which can be very confusing for students.

3.3 Chemical subjects in lower secondary education

3.3.1 Black box

Introducing chemistry to kids of age 12 to 14 is complicated. For them, it is difficult to enter an imaginary world of atoms and molecules, where specific rules apply and a world that cannot be made visible. A world that cannot be seen; only experiments give indirect indication of properties that exist in this microworld.

One of the ways in which students can get some idea of the way information can be obtained of something that is not visible is the following experiment.

If you build a small rectangular box of an opaque material like wood, triplex or similar, you can include an object like a small ball, a nut, a bolt, a piece of rubber eraser or things like that in the box before closing the box (Figure 3.6). Painting the box black will yield a black box.

Students can be asked to make a drawing of what they think is inside the box. They can also be asked to give some idea of the material the object is made of.

You will be surprised how accurate the descriptions can be (Yayon & Scherz, 2008).

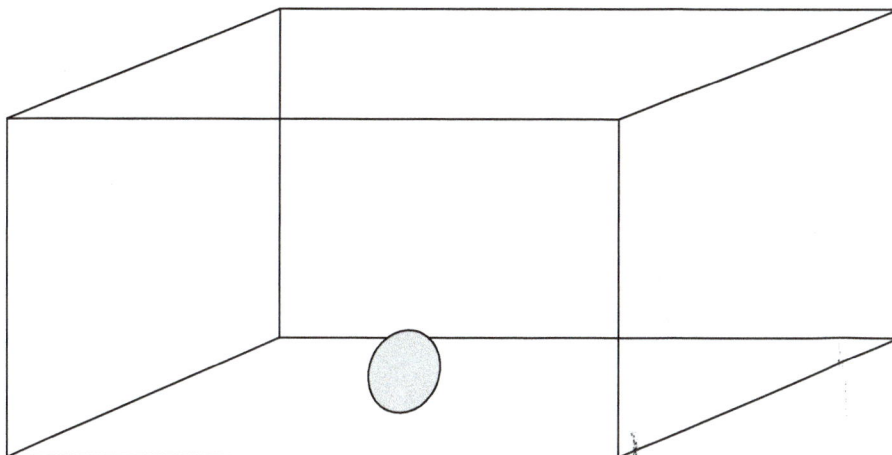

Figure 3.6: Rectangular box, containing a small ball.

Describing experiments from which information can be inferred about the atomic scale should help students understand that world a bit better. One of the most difficult concepts to introduce to students is the idea of electrical charge. It is not something they have experienced before. They are familiar with electrical current as a source of energy, but electrical charge is very often a new completely abstract concept. As his concept plays a role in physics as well, it is important to discuss with your colleagues exactly how you are going to introduce this concept. Any misconceptions students develop will be because of the way the concept of charge is introduced to them. Charge is often linked to static electricity, using a van der Graaf generator or a Wimshurst apparatus to generate sparks (Figures 3.7 and 3.8).

Relating the concept of charge, as introduced to these macroscopical phenomena and to the atomic/molecular level is not easy for students. Especially subtle things like hydrogen bridges, polarity and so on.

3.3.2 Properties and the concept of a substance

One of the first concepts students should learn is the concept of properties of a substance. This can be discussed at the macroscopic level, making it a lot easier. Again, the concept of a pure substance is not something that will be familiar for students. Dubbing a particular material, a substance is new for them. Let alone the difference between a pure substance and a mixture of substances. Especially since the word "pure" means something else for a chemist than what it means in common language. Consider pure air, or pure water, for example.

To introduce the concept of properties of a substance, experiments are very often used. In most textbooks, an experiment is described in which students are

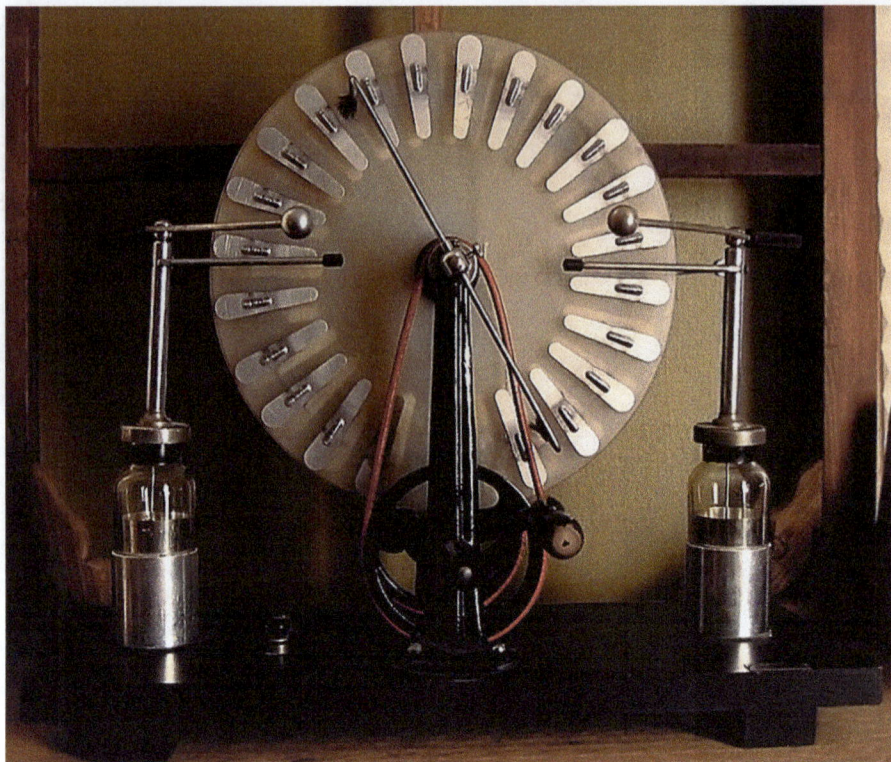

Figure 3.7: Wimshurst generator (wikipedia).

given a set of maybe ten to fifteen different substances. They are asked to give as many properties of each substance as possible. At first, this is done at the macroscopical level, with a focus on identifying exactly what properties are. Properties like smell, texture and color will make it possible for students to differentiate between sugar and salt and between ammonia and water. It must be made clear that "amount," "temperature" and "shape", for example, are not properties of a substance.

Once the concept of properties is clear to students they can move to the next step, that is, studying the difference between pure substances and mixtures (Figure 3.9).

Separation techniques

The relationships between these concepts, or how to get from one to the other, are of course important as well. The link between mixtures and pure substances are separation methods.

Figure 3.8: A boy touching the high voltage terminal of a Van de Graaff generator at a science museum (wikipedia).

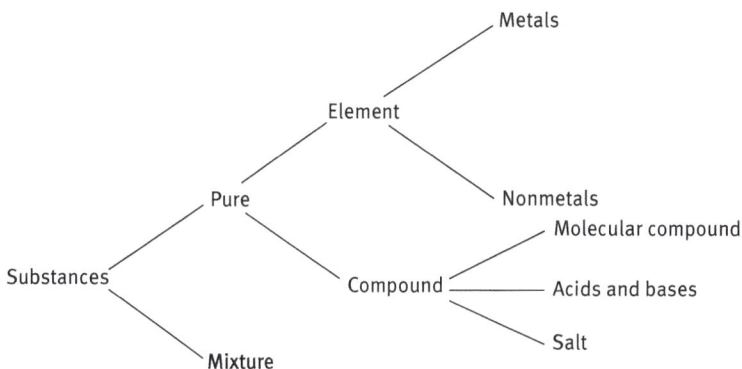

Figure 3.9: Overview of concepts for an introduction to chemistry.

Very often the first or second chapter in a book introduces separation methods.

Before going into separation methods, it is important to show students why separation is such an important method. One of the ways to do that is to relate to the UN development goals. The UN has adopted 17 sustainable development goals that constitute great challenges for the upcoming 20–50 years (Figure 3.10).

Figure 3.10: UN sustainable development goals.

One of the UN sustainable development goals is clean water and sanitation. Obviously, the production of clean water involves the use of several separation techniques. Students may be asked to find out the ways clean drinking water is produced all over the world. Filtration, precipitation, distillation and adsorption are bound to come up. Most western countries use surface water, or ground water that is treated in a purification plant, involving filtration, coagulation and precipitation. In the Arab world, seawater is distilled to produce drinking water. Small-scale purification of water often involves distillation.

By linking separation to such a societal issue, the concepts receive a much larger impact than when you just introduce them as one of the tools in chemistry, which of course they are.

Extraction and evaporation are two methods that are introduced at this point as well. They can easily be linked to be responsible for consumption and production. A perfect example is the different ways in which coffee can be made. To make coffee in the first place, extraction is needed of the ground coffee beans. For instant coffee, evaporation is needed, either by freeze drying or by another technique. For caffeine-free coffee, extraction is needed. Linking these processes to everyday life gives the techniques meaning to the students.

Later, they can be linked to uses in chemistry.

Chromatography is a technique that is often used to identify compounds of a mixture. It is not used very often for purification purposes. It is questionable whether it should be introduced here or at some later stage, when it is more appropriate or can be linked to its use more clearly. Especially since paper chromatography, which is often introduced at this stage, is not used very much in chemistry in any way. Gas chromatography and HPLC are techniques used widely, but are complicated to be introduced at this level. It can easily be left out at this stage.

From the description provided previously, it is clear that up to this point no explanation at the microlevel has been needed, to learn about the separation techniques. If you want to describe the atomic/molecular background of the separation techniques, an introduction to the molecular model is needed. Again, the molecular model is something that is needed in physics as well. Very often students have discussed aggregation states in physics and have heard about molecules. It is extremely important that the molecular model used in physics is the same as that used in chemistry or biology. Consultation with your colleagues on this point is needed, just like consultation is needed about the way electrical charge is introduced.

Carry out assignment 3.1

Chemical reactions

The link between elements and compounds is decomposition and synthesis. Since these are special cases of chemical reactions, the concept of a chemical reaction needs to be introduced.

The concept of a chemical reaction or a chemical change is not very complicated. Students will easily accept the definition of a chemical reaction as a process in which before the reaction starts you have substances and after the reaction you have different substances. There are plenty of examples in which you demonstrate. One of the oldest is mixing iron and sulfur, and showing that the mixture still has magnetic properties, for example. When heating the mix, an obvious reaction takes place, resulting in a new substance. Other demonstrations can be found easily on you tube or on the Internet. See, for example, https://youtu.be/WPmYlBk_utE . The reactions performed are burning of magnesium, the heating of baking soda, the reaction between copper sulfate and an iron nail and the reaction between potassium chloride and silver nitrate; these are classical examples of chemical reactions. They have no daily life association.

This is actually a good time to start discussing the language and symbols that chemists use for describing chemical reactions. The substances before the reaction are called reactants, the substances after the reaction are called products.

The easiest way to describe a reaction similar to the one mentioned above would be:

A metal burns with a white flame, yielding a white powdery substance
Identifying the metal as magnesium and the powder as magnesium oxide leads to the description:

Magnesium reacts with oxygen to yield magnesium oxide, which is more specific

Using symbols, it might read as follows:

$Mg + O \rightarrow MgO$

Before you can go further and into more details, you need to make a link to the microscale, introducing Dalton's atomic model. The atomic model of Dalton, in which atoms are considered to be solid spheres is adequate at this level. With this model, the difference between elements and compounds can be explained, chemical formulas can be derived, a chemical reaction can be balanced and so on.

Atoms have certain rules:
– Molecules are built up from smaller particles called atoms
– A molecule may contain between one and many atoms
– Atoms rearrange in new molecules during a chemical reaction
– Atoms cannot be destroyed or created in a chemical reaction
– There are a bit more than 100 different types of atoms
– There is an infinite number of molecules built up from atoms
– For each type of atom, a specific symbol is used, normally derived from its Latin name.
– Molecules that contain only one type of atoms are called non-decomposable molecules or elements

Very often in textbooks these rules are garbled and are not presented in a coherent manner.

There is quite some discussion between chemists about the definition of the concept of an element.

In IUPAC's goldbook (https://goldbook.iupac.org/html/C/C01022.html), which is normally used as a standard, an element is defined as follows:

1. *A species of atoms: all atoms with the same number of protons in the atomic nucleus*
2. *A pure chemical substance composed of atoms with the same number of protons in the atomic nucleus. Sometimes this concept is called the elementary substance as distinct from the chemical element as defined under 1, but mostly the term chemical element is used for both concepts.*

This means that both the ideas of what an element is can be used.

In order to describe a chemical reaction in terms of a balanced chemical equation, the symbols for the elements are needed as well as symbols for compounds. Introducing the periodic table in order to differentiate between metals and nonmetals seems logical.

Element bingo (see Figure 3.11) is one way to help students memorize the symbols. Have the students make up a bingo card with either 9 or 16 elements. They can either use only symbols or only names. You read out the names or the symbols of the elements. 3 or 4 elements in line would be bingo.

Ag	Na	Cl
O	H	Pt
Ca	C	Sn

Aluminum	Fluorine	Helium
Nitrogen	Barium	Lithium
Lead	Boron	Gold

Figure 3.11: Example of 3 × 3 bingo cards.

Students find the use of symbols confusing, and difficult to learn. The symbols are difficult to remember since they seem arbitrary. Linking the symbols to the Latin names of the elements may somewhat help. In addition, the way chemists use a formula to indicate the atomic composition of a substance is arbitrary and needs practice by the students.

An important aspect of the introduction of the use of symbols and equations is the difference between the micro and the macro stage. Either you are dealing with (single) atoms and molecules, or you are dealing with substances. This difference should be made explicit to the students (Table 3.1).

Table 3.1: Steps in balancing a chemical equation.

Macro	A silver gray metal burns with a white flame, and yields a white powder				
Macro	Magnesium	And	Oxygen	yield	Magnesium oxide
Macro	Mg	+	O	→	Mg,O
Micro	Mg	+	O_2	→	MgO
Micro	$2\ Mg$	+	O_2	→	$2\ MgO$
Macro	$2\ Mg(s)$	+	$O_2(g)$	→	$2\ MgO(s)$

There is one more thing that needs to be considered when discussing chemical reactions. That is, the number of atoms and molecules involved. Students feel that one million atoms and molecules are a huge number of molecules. When discussing organic reactions, or equilibrium at a later stage, the enormous number of molecules is a very important aspect. Students tend to think in terms of single molecules and atoms.

Carry out assignment 3.2

The good thing about the concept of chemical reaction and the difference between elements and compounds is that they can be discussed first at the macroscopical level. Introducing molecules and atoms is not necessary at first, which makes the concepts a lot easier. On the other hand, the link to the atomic and molecular scale gives a chemistry concept.

At a certain point the link to the microscale is needed to explain phenomena, specifically, when different types of compounds are discussed.

A white substance that dissolves in water and become invisible is baffling. Even though describing the difference between a sugar and a salt solution on a macroscopic scale is not a big deal, it is not very obvious. Why would you be interested in the fact that a salt solution conducts an electric current and a sugar solution does not. Describing this at the micro level is a lot more complicated. Especially since at the micro level, subatomic particles like electrons, protons and neutrons are needed to fully understand the difference.

That is why this difference, as well as the introduction of ions, is often postponed until the first chemistry course in upper secondary education.

Carry out assignment 3.3

Other subjects

Once the concepts of Figure 3.9 are clear to students, other concepts could be studied. In the past, specific types of chemical reactions were studied. Combustion is very often introduced, including the introduction of organic chemistry, with at least the alkanes and most often ethanol. The combustion reactions were often used to practice chemical calculations.

In the last 10 to 15 years, other subjects have been introduced that have more meaning to students. Since this introduction to chemistry is often the only chemistry for quite a few students, it is important to demonstrate the important role chemistry plays in society, both in a positive way as well as in the negative way.

It would seem logical to use the sustainable development goals from Figure 3.10 as a guideline for other chemical subjects to introduce to students. In Table 3.2, a number of possible links are given.

There are many possibilities for linking introductory chemistry to these sustainable development goals. Table 3.2 is just a first inventory. It is important to demonstrate to the students the role chemistry could play in possible solutions for reaching the UN development goals. More important is to show students the way in which they can contribute to achieving the sustainable development goals, giving them ownership of the problem.

Table 3.2: Some suggested links between sustainable development goals and chemistry subjects.

Goal	Chemistry
1: No poverty	No obvious link
2: Zero hunger	Link to fertilizer, phosphate and nitrate cycle, pesticides, fungicides, genetic manipulation
3: Good health and well-being for people	Developments in pharmacy, health care, use of antibiotics in cattle breeding
4: Quality education	No obvious link
5: Gender equality	No obvious link
6: Clean water and sanitation	Drinking water preparation, pollution
7: Affordable and clean energy	Different types of energy, fossil fuels
8: Decent work and economic growth	No direct link
9: Industry, innovation and infrastructure	Safety in chemical industry
10: Reducing inequalities	No direct link
11: Sustainable cities and communities	Several links, cradle-to-cradle design
12: Responsible consumption and production	Cycles of elements
13: Climate action	Ecosystems, systems thinking
14: Life Below Water	Pollution, plastics in the ocean
15: Life on land	Pollution, use of pesticides
16: Peace, justice and strong institutions	No obvious link
17: Partnerships for the goals	No obvious link

The subjects chosen for this second part of an introductory chemistry course may differ for different groups of students. Students who prefer to continue taking chemistry courses in upper secondary school can be offered an alternative program than those students dropping chemistry in the next stage.

Carry out assignment 3.4

3.4 Difficult concepts and alternate conceptions

An introductory course as described previously has some concepts that may give rise to misconceptions or rather alternative conceptions. You have to be careful in explaining these difficult concepts. They have to do with representations of the microscale. This relation between the macroscale and the microscale is challenging for students. The molecular model as well as the Dalton's atomic model are not always clearly understood. Students tend to mix up the macro- and microscale.

Atoms are often depicted as small spheres, having a certain color. Based on these illustrations, students sometime think that a sulfur atom is yellow colored and a carbon atom is colored black. Connected to these illustrations are the ball and stick models that are used for modeling molecular compounds. Students tend to think that atoms are actually connected by some small sticks.

The obvious solution is to make clear that these illustrations are a model of reality. The concept of a model is not simple for students. Why we use models and what the value of using models is, is not always clear to students.

You should be able to find enough articles about misconceptions.

Carry out Assignment 3.5

The meaning of a balanced chemical equation, for example, is one of the more difficult ideas to grasp for students (Kimberlin & Yezierski, 2016). Several concepts needed to write a balanced chemical equation need to be learned and understood before starting to teach balancing an equation. Students may learn to use the algorithm to balance an equation, but understanding what it means is something else (Nurrenbern & Pickering, 1987).

Underlying the concept of balancing an equation is the use of formulas to describe substances. Understanding the meaning of all the details of $2\,Al_2O_3$ and the difference with $2\,Al_2O_3(s)$ is quite challenging for most students.

The concept of limiting reagent, which is related to balancing an equation is equally hard to understand for students.

As the concept of a balanced chemical equation is essential for any further work; for example, in stoichiochemistry, it is very important students have a good understanding of the concept.

One of the main problems a teacher can run into at this stage is assessing the learning process his or her students have made. It may seem as if students have understood concepts completely, while all they have done is learned an algorithm and focused on passing a test. Assessing the deep level of knowledge students have is quite a challenge. Knowledge linked to problem-solving can be described at different levels (Jong, 1996); each level is characterized by certain actions. Four types of knowledge are discerned:

- Situational knowledge, basic knowledge about a specific situation, being able to solve problems within that situation
- Conceptual knowledge, aware of more basic principles, able to use formulas and symbols
- Procedural knowledge, conscious choice of algorithms, use in other contexts
- Strategic knowledge, overall overview, linking to other situations, ability to integrate principles with other principles

Assessment should focus on these levels of knowledge to determine the level of knowledge a student has acquired. In Chapter 4, knowledge will be discussed in more detail. In Chapter 8, assessment will be discussed more deeply.

Assignment 3.1 Introducing charge

Discuss with your physics colleagues the way they introduce the concept of charge and the molecular model. Identify the way they introduce the concepts and the words they use to describe phenomena.

What are the learning goals they want to achieve? How will this influence the way you introduce these concepts to your own students?

Make a short presentation for your fellow students, and compare what they have found.

Assignment 3.2 Balancing equations

Design an activity to introduce balancing equations, using LEGO, or a similar material.

Look up a textbook that was used around 2000 to introduce chemistry in secondary school. Link the chapters in that book to Figure 3.2.2.1.

Then take the textbook that is used now, or use something like "chemistry for the Twenty-first century" (Borley et al., 2016) or "active chemistry"(Eisenkraft & Freebury, 2003).

Link the chapters in one of these to Figure 3.2.2.1.

Observe and describe the difference in which concepts are introduced to the students.

Prepare a short presentation in which you highlight the differences.

Give the presentation to your peer group, or to the other teachers in your department.

Assignment 3.3 Using a context

Design an introduction of one or two lessons for one of the concepts or group of concepts, using a context chosen by yourself.

Think, for example, about the ideas of a limiting reagent related to a chemical reaction, which may give rise to discussions on sustainability.

Another idea is to use a hydrogen powered car to introduce some of these concepts.

Indicate the knowledge the students should have before they start this lesson series. Indicate as well what the students should have learned at the end of the lesson series.

Present your lesson series to your coach and your peers. Use the feedback you receive to improve the lesson series. Finally, carry out the lessons with one or two of your classes.

Write a short report to be added to the material in which you analyze what went well, what should be changed the next time you use this lesson series.

Assignment 3.4 peer lesson design

Step 1 Together with your coach, fellow students design a lesson plan for the introduction of balancing a chemical equation. Students should be aware of Dalton's atomic model and the molecular model. They should know the symbols of the most used elements.

Step 2 Let one of you carry out the first lesson of the designed lesson plan, while you and others are observing in the classroom, with a special focus on the learning process of the students.

Step 3 Evaluate and discuss the effect of the lesson. If necessary, adapt the lesson.

Step 4 Let one of the others carry it out now, observed by the rest.

Step 5 Evaluate and discuss once more. If needed, adapt the rest of the lesson plan.

Step 6 Design an assessment that can be used to determine conceptual understanding versus rote learning of problem-solving. If needed, use Nurrenbern's article (Nurrenbern & Pickering, 1987) as a reference.

Step 7 Evaluate the assessment results from you and your fellow students, and discuss the results.

4 The process of learning

4.1 Introduction

Learning is a process that begins at birth and lasts lifelong. Adapting to your environment in a productive way is the basis of learning. Parents, elders and peers help you find your way in society. In early societies, specialists in hunting and tool building evolved. Often master–apprentice relationships developed, in which knowledge was passed on from one individual to another. Schools as such did not exist.

In early societies like Egypt and India, schooling was focused on writing and religion. In China (Chan, 2006), imperial schools emerged in which people passed exams in order to become state officials. The program was based on Confucianism and taught six classic arts, among which were rites, calligraphy and mathematics. It had three levels of degrees that could be chosen, comparable to modern degrees like bachelor's, master's and PhD.

When people began to settle down in villages and towns, relationships between people changed, and society became hierarchal. People who did not own land became dependent on those who did. Schooling was reserved for the well-off people. In these rich families, writing and arithmetic were taught by private tutors. In ancient Greece and Rome, schools were private institutions that most people tried to attend for at least two years, but the cost was high.

In the sixteenth and seventeenth centuries, schools and universities began to emerge in Europe, where Latin, the universal language of Europe, was taught. Pursuing mathematics and philosophy, as well as medicine had already begun; texts from Greek, Arab and Roman sources were taught. In Massachusetts, a religious school was established in the mid-seventeenth century.

It was not until the nineteenth century that schooling became widespread, and that governments began to play a role in education. It lasted until the twentieth century before child labor was abolished and schooling became generally available (Gray, 2008).

As depicted in Figure 4.1, primary education focused on basic skills that were called the three R's: Reading, wRiting and aRithmetic. The importance of these skills is evident for any further education. Secondary education takes over from the master–apprentice type of learning to a more general vocational training. This broader background prepares individuals for a different role in the society.

In the second half of the nineteenth century the need for human capital increased, and secondary schools emerged and became more general. The role of education and schooling in adding value to human capital is discussed clearly by the Nobel Laureate Theodore Schultz (Schultz, 1989). Economic development is clearly linked to education and schooling.

https://doi.org/10.1515/9783110569629-004

Figure 4.1: Counting frames in the classroom.

In present times, these roles of education have become clearer. In primary education, history and geography have found a place, giving the students a broader sense of the society and their responsibilities toward the society, and preparing them for further education.

Secondary education has diversified in several streams. One is vocational, focusing on jobs like construction or mechanical where skills play an important role. Another stream provides a general education that can be used as the starting point for further education, for example, in administration. The last stream prepares for higher education. A more general aim of present-day secondary schools is the development of citizenship competencies (Geboers, Geijsel, Admiraal, Jorgensen, & ten Dam, 2015) needed to cope with daily life in a complex society.

Children between the age of 12 and 18 experience major changes in their development, both physical and mental. Secondary education helps them cope with these developments and provides them with a safe background for learning.

In secondary education, students on one hand need to learn how to learn, how to acquire information and how to manage and allocate time for different activities. On the other hand, they need to increase their basic knowledge. Arithmetic is about learning algebra and geometry. Learning to read is not only comprehensive reading, but also learning other languages to deepen the knowledge about the structure of the language. It is also very important to learn about other cultures. Learning how to write and use different structures and types of documents for clear, logical and concise formulation is vital.

At the same time, they go through puberty and adolescence, experiencing emotional and physical changes. They should learn how to behave in the society, as well as develop their own identity. All these aspects need to be blended with their learning at school. School is as much a place for social development as it is for learning specific subjects. After secondary school, students need to have some idea about their future role in society and what they want to achieve in their lives.

For science students, the challenge is to inquire and investigate about things they experience in their daily life. They should be curious about the phenomena they accept as "normal." Why does light turn on when you flip the light switch? Why can you recharge certain batteries, and what is the problem if you try to recharge batteries that are not rechargeable? Teaching students the applications of science in different fields should stimulate their curiosity. Science education should stimulate inquiry.

In higher or upper secondary education, the level of learning is high and its process is different. One of the important factors in higher secondary education is to stimulate the interest of students. At some point in time they have to find out which subjects they find interesting enough to pursue further. The science lessons need to demonstrate what going further in the field of science entails, what the challenges are and, more important, what fun in science is.

It is so important to demonstrate the role science plays in the society, especially in the twenty-first century, where sustainability and sustainable development goals (Nations, 2016) are the major challenges for the society. New scientists are needed to work on these sustainable development goals.

4.2 Early research and thoughts about teaching and learning

Research and thoughts about learning have developed through the centuries. Controversies in learning style, for example, between Plato and Aristotle, have been well described (Hein, 1975). With the development of schools, and governments taking the responsibility of education, research about teaching and learning started up on a larger scale.

Pioneers involved in the development of the education system focused mainly on primary education. Maria Montessori (Montessori, 1912) was inspired by the work of a French scientist, Itard, and read the learning of "the savage of Aveyron." She developed the "Montessori" method for teaching children. This method is based on what she called sensitive periods. These are the periods in which a child is sensitive for learning a specific subject. It is the teacher's responsibility to recognize those periods and offer the child the appropriate material to study. The material is designed in such a way that students can work with it independently. Children make large leaps of progress this way. Montessori schools based on her methodology are still functioning all over the world (Figure 4.2).

Figure 4.2: Montessori materials – pink tower and broad stairs.

Peter Petersen of Jena, Germany (Hooijmaaijers, 2000), developed a methodology in which children worked in social groups. This program is based on loose parts and is then joined together by groups of children (figure 4.3).

In France, Célestin Freinet (French Embassy NY French, Cultural Services, 1971) worked on a program in which a printing press was the central item in the classroom. Students printed their own study materials. At the beginning of each week, the goals for that week were set in a class meeting (Figure 4.4).

Tasks were allocated to the students and work commenced. Dorothy Parkhurst (van der Ploeg, 2014) worked with the weekly tasks of individual students (figure 4.5).

Figure 4.3: Jenaplan school in Jena.

Figure 4.4: Printing press in a Freinet school.

Figure 4.5: First Dalton school.

The common factor for all these methodologies of teaching is that children are the center of education, involved in their own process of learning and at the same time forming an individual learning trajectory. This trend of an open education system has evolved remarkably (Hein, 1975).

Rudolf Steiner started the so-called Waldorf schools (Prescott, 1999) in 1919. On the one hand, education gives children the freedom to develop themselves, on the other hand, education is heavily linked to the ideas and philosophy of Steiner. These schools exist all over the world and are popular among a large number of people; however, some criticize Steiner's rigid ideas and mysticism. Artwork plays an important role in Waldorf schools; see Figure 4.6.

4.3 Piaget

Learning to interact with your environment with the help of your parents, family and peers passes through multiple stages. The way experiences are interpreted and remembered pass through stages in which the insight into observed phenomena deepens.

Figure 4.6: Artwork from a Waldorf school.

Piaget (Figure 4.7) indicates that there are four periods of intellectual development taking place in an individual. This development is hierarchal and to an extent related with age. You must pass through the previous stage before going to the next.

The first stage, called the sensorimotor phase, takes place during the first two years after birth. One of the developments is that of object permanence. In early stages a child does not remember an object. When it is hidden beneath a blanket, for example, it is simply not there. After 6 to 12 months, a child remembers where an object is.

The preoperational stage, between the age of two and seven, is characterized by a further development of motor skills when a child develops a sense of identity, remembers his name and so on. One of the tell-tale signs of this phase is the concept of quantity. When a child confronts two short, wide beakers filled with lemonade of

Figure 4.7: Piaget and Queen Juliana of the Netherlands at the occasion of Piaget winning the Erasmus prize.

the same level, he or she will indicate that the beakers contain the same quantity of lemonade. When one beaker is poured in a tall, narrow glass in which the water level rises higher, he or she will indicate that the tall glass contains more liquid.

When two rows of six evenly spaced coins are made, a child will answer that they contain the same number of coins. When the distance between the coins in one row is increased, the child will indicate that that row contains more coins.

The third phase, called the concrete operational phase, lasts throughout primary school and beyond. In this stage a child learns to classify things, develops reasoning skills, understand conservation laws and is able to follow algorithms.

The fourth phase, the formal operational stage, is distinguished by the possibility of abstract reasoning. Objects do not need to be real, but can be imaginary. Logical reasoning may be applied. At some time during the secondary school years, this phase evolves.

There are some simple tests to check whether or not a student has reached this stage. One of the most-asked questions is:

If Kelly is taller than Aly and Aly is taller than Joe, who is the tallest?

Another question is:

If you could have a third eye, where would you like to have it?

In science and mathematics, students in the concrete stage will try to solve a problem by trial and error. Basically, they will try what they learned first, if that does not work they try another way of solving the problem. They might have difficulty in improvising.

Someone in the formal operational stage should be able to look at the problem and analyze it before trying out a solution.

Especially in problems that require sequential steps in solving them, someone who has not yet reached the formal operational stage will experience difficulties.

Abstract thinking comes with the formal operational stage. The concept of atoms and molecules interacting with each other according to certain rules is an example of reasoning that sometimes becomes difficult. Students not yet fully in the formal operational stage will experience problems understanding what they are dealing with, and will fall back on macroscopic similarities that they can understand.

Fortunately, the transition from one stage into another is a continuous process. It can be stimulated and learned by students. During their education, it is possible to train the students in abstract thinking.

Another factor is that individuals do not demonstrate formal operations in all tasks (Day, 1981). In some tasks they may develop formal operations while they may not use them in others. This also indicates that learning plays an important role in the use of formal operations. Students use formal operations in situations they are familiar with, of which they know more. In new situations where they have less experience, they fall back on concrete operational skills.

For secondary school education, it is important to know and ascertain that students have attained the formal operational stage for the tasks and problems that they are confronted with. The training and teaching of students should take the level of students into account, so that they can be trained to develop themselves. On the other hand, it is important to note that children below the age of 12 rarely develop formal operational thinking skills (Day, 1981).

4.4 Vygotsky

Lev Vygotsky was a Russian psychologist, who died in 1934 when he was only 38 years old. He worked on developmental psychology like Piaget, and at about the same time as Piaget. His work was not known in the western world until about 1980, when a translation of an accidentally discovered work was published (Vygotsky, 1978) (Figure 4.8).

When Piaget defined stages of development, Vygotsky focused on the learning process itself (Mcleod, 2014). Children learn through interaction with others, such as their parents, siblings, friends and other families. They are confronted with different situations and learn to cope with the situations normally with help from others.

Language plays an important role in learning as it is the vehicle most often used for instructions, although more concrete actions can also be used. At a

Figure 4.8: Lev Vygotsky (1934).

certain age, children begin to internalize language and producing verbal thoughts. Vygotsky indicates that internalization of language leads to cognitive development.

Vygotsky sees this as the construction of knowledge in an individual. This construction of knowledge is discrete, as it takes place in the mind of an individual. But this construction is determined by the sociocultural environment with which a child is confronted. Like Piaget, Vygotsky feels children are curious and actively involved in their own learning, which tie in with the active periods of Montessori. On the other hand, Vygotsky considers that thinking is controlled by social interactions with others, which ties in with the ideas of Petersen and Freinet. Children try to understand verbal and other instructions from their tutors, internalizes them in their mind and use them to regulate their actions.

Vygotsky defined a zone of proximal development for the learner, which the more knowledgeable other, most often a teacher or peer, needs to be aware of. Vygotsky indicates there are tasks that a child cannot do on his or her own, but is able to with the help and guidance of someone who is more knowledgeable. This allows the child to develop skills that can be used later to perform the task on his or her own. This also means that this zone has continuous changing levels (Figure 4.9).

Language is extremely important in the learning process, as language is used for giving and receiving instructions. In case of social languages, a child uses the

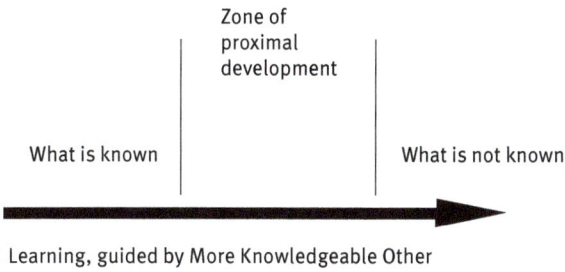

Figure 4.9: Zone of proximal development.

language and words of his or her own, demonstrating what he or she has learned. Interaction with peers helps a child internalize concepts in his or her own mind. A child or student can discuss concepts only when he or she has internalized them. Creating settings in the classroom in which students can interact with each other are extremely helpful for their learning process (E. G. Cohen, 1994).

This theory of knowledge construction is the base and background of new teaching methodologies. One of the important conclusions is that new knowledge is linked to already existing knowledge. Concepts are constantly adapted based on the information received. A beautiful example of this process is given in a children's book called "Fish is Fish" written by Leo Lionni (Lionni, 1970). It is the story of a tadpole and a goldfish that grow up together in a pond. They become close friends. But after a while the tadpole grows legs, and ultimately becomes a little frog. He decides to hop out of the pond and explore the outside world. After a few days he comes back. He then describes birds, cows and people to the goldfish. Leo Lionni draws the mental images of the fish with the described things. Sure enough the cow is black and white, has horns and has a pink udder, but basically it is still a fish. He has all the described details, but on the whole the bird is still a fish with wings attached, so that it can fly. You can find these images on the web.

The important message is that these preconceptions interfere with the construction of correct concepts. This is illustrated with a video produced based on a research in Harvard, called "Minds of Our Own" (Astrophysics, 1997) in which engineering graduates from MIT (the Massachusets Institute of Technology) ask questions about photosynthesis or using a battery to light up a bulb. In the video it becomes clear that despite what students learned about closed loops and photosynthesis, they tend to fall back on previous conceptions. Learning is a continuous process that needs to be nurtured constantly.

Again, it needs to be noted that assessing the ideas students have is absolutely vital to monitor their learning process.

4.5 Processing information

Vygotsky and Piaget developed theories about the way information is stored and built up in what is called the long-term memory. Knowledge can be recalled from this long-term memory into the working memory. Others have thought about the way information is processed. Baddeley developed ideas about the relation between working memory and long-term memory (Johnstone, 1997). Figure 4.10 depicts this relationship.

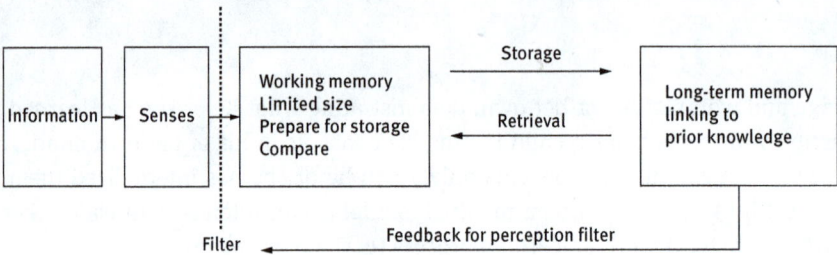

Figure 4.10: Baddeley's model. From Johnstone (1997).

In this model, information is received through the senses, filtered by a perception filter, stored in the short-term memory or working memory, then compared to the prior knowledge in the long-term memory and is either stored or deleted.

Baddeley partitioned the long-term memory in a visual-based memory and a phonological word–based memory (Mayer, 2009) (Figure 4.11).

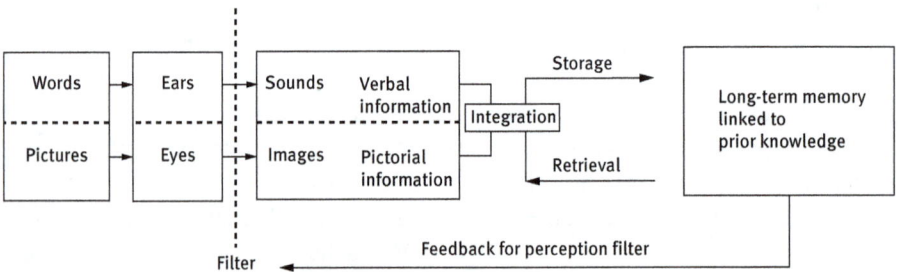

Figure 4.11: Extended model of Baddeley. From Mayer (2009).

For learning, it is important that the information received through words and pictures is aligned. When that is not the case, information becomes mingled and is much harder to store (Mayer, 2009).

The process of information storage is complex, and is influenced by a lot of factors. Concentration is one of them. Not all information received is stored in the short-term memory. Redundant information, for example, is observed but not registered.

The process is also individual. The knowledge stored in the long-term memory influences the perception filter. The already-existing knowledge plays an important role in the filter.

In an experiment in 1945 (Groot, 1946), people were asked to look at a chess board for 30 s or so and to reproduce the board after a short while. Most people were able to place five or six pieces back in the right place. An example is given in Figure 4.12.

Figure 4.12: Chess board, Bogo-Indian defense after move 16.

When the board was shown to chessmasters, they were able to reproduce all the position because they recognized it from the chess theory.

When chessmasters were shown positions not related to theory, they also fell back to five or six pieces.

This indicates a maximum size of the short-term memory of about six items. In this case, it would be the short-term memory related to images. For words, similar experiments can be performed. Again, there it seems that the number of spaces in short-term memory is four to six (Figure 4.13).

The size of short-term memory also has an influence on the performance of students on solving problems. On the horizontal axis, the sum of the number of pieces of information needed to solve the problem is indicated. On the vertical axis, the percentage of students able to solve the problems correctly is given. This number drops dramatically after five variables.

Succes rate of solving problems

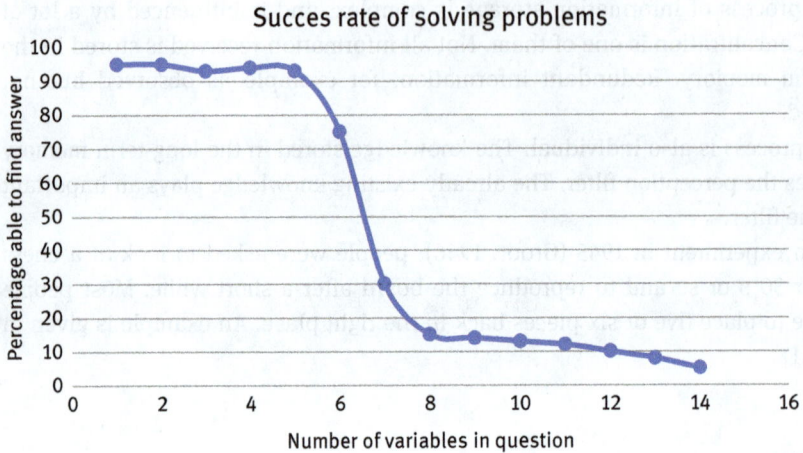

Figure 4.13: Success rate of solving complex problems. From Johnstone (1997).

Assignment 4.1

There are more individual differences in which information is processed. One of these is the preference for one of the senses. Some people prefer visual input, others prefer auditive information. A third group likes what is called kinesthetic input that can be provided, for example, by writing. Most people have a first preference and a second preference, and for some there is no preference. In neurolinguistic programming (Carey, Churches, Hutchinson, Jones, & Tosey, 2010), these differences are studied.

Visual learners like diagrams and symbolic presentations. Auditory learners engage better in a lecture session, reading out loud or talking to themselves. Kinesthetic learners do well in hands-on environment, highlight texts and enjoying role-play scenarios.

It is fairly easy to determine someone's preference. A simple exercise in listening reveals the use of specific words relating to a preferred sensory input.

Assignment 4.2

Some preferences can be considered when designing lesson activities. A Power-Point presentation will cater to visual learners, while talking and discussion will help auditory students. Laboratories and experiments will help kinesthetic pupils.

4.6 Learning styles

It is clear that learning is an individual process with individual results. This means teaching needs to cater to different modes of learning. There are more possible aspects of learning. In the 1980s, Kolb described his cycle of experiential learning (Konak, Clark, & Nasereddin, 2014).

Table 4.1 lists the four stages of learning described by Kolb.

Table 4.1: Steps in Kolb's learning cycle.

Concrete experiences	A learner experiments with new ideas, relating them to his or her own experience, he or she becomes personally involved.
Reflective observations	A learner looks at the new ideas from different points of view and thinks carefully about the new information.
Abstract conceptualization	A learner tries to organize the information in logical blocks and to link it to existing theories in order to understand better.
Active experimentation	A learner applies the new information directly in situations to solve real problems, he or she takes action with the new material.

The cycle is often described by a figure in which the steps follow each other (Figure 4.14).

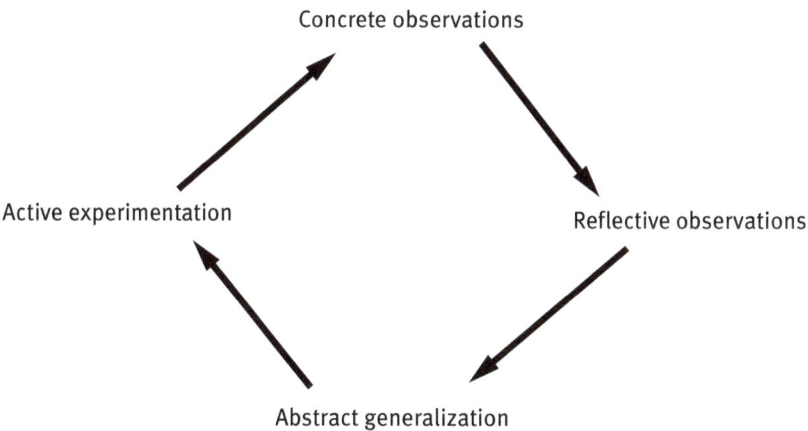

Figure 4.14: Kolb's experiential learning cycle.

According to Kolb, each step in the learning cycle must be followed completely in order to learn new information correctly.

Kolb indicated that learners can be divided in four learner categories based on these learning steps (Kuri, 2000). Learners have preferences about the way they try to learn new material.

In Table 4.2, these different types are listed. They generally fall between two steps in the learning cycle.

Table 4.2: Kolb's learning styles.

Accommodating style	Between concrete experience and active experimentation. People learn from hands-on experience, prefer to work together.
Diverging style	Between concrete experience and reflective observation. People look at concrete situations from different points of view, good at brainstorming, like to work together.
Assimilating style	Between abstract conceptualization and reflective observation, people like to work alone, look at a wide range of information, put it in a logical form.
Converging style	Between abstract conceptualization and active experimentation. People like to work alone, prefer technical tasks and theories, applying theory to practice.

Students should be confronted with these different learning styles, and learn to follow other learning styles. One of the ways to accomplish this is to group students according to their learning styles and let them work together on larger tasks (Figure 4.15).

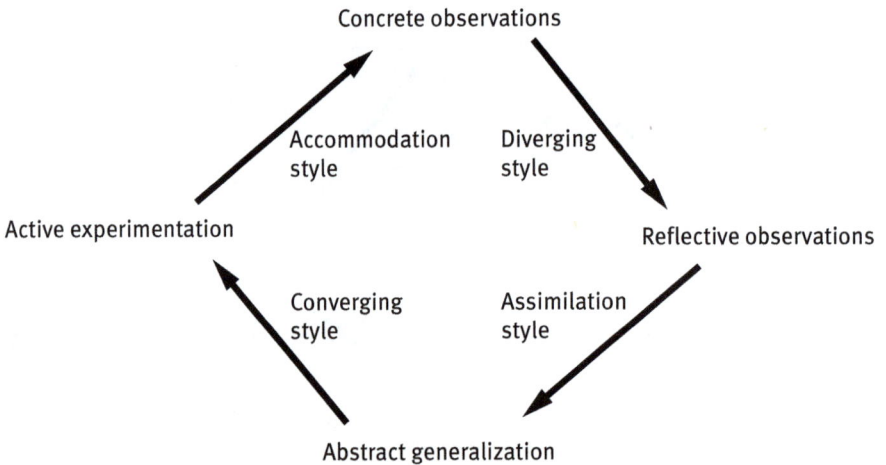

Concrete observations

Accommodation style Diverging style

Active experimentation Reflective observations

Converging style Assimilation style

Abstract generalization

Figure 4.15: Learning styles fitted to learning cycle.

It is not very easy to adapt to these different learning styles in teaching. The most important aspect is that as a teacher you are aware of these learning styles.

In your teaching you can demonstrate the different approaches to work on a problem. When students are aware of their own preferred learning style, it will be easier for them to look at new material from a different perspective and go through the complete learning cycle.

4.7 Multiple intelligences

Intelligence is a difficult concept. It gives some information about the learning capabilities of students. Intelligence tests will yield an intelligence quotient (IQ), which gives some information about the relative position of an individual as a whole. These tests have been standardized, and are normally administered through professionals in psychology or pedagogy. For a teacher, it is important to know the IQ of a student to determine whether the student is gifted or not. Gifted students often have different ways of learning. Determining whether or not a student is gifted can be important in offering him or her alternative ways of learning fitting to his or her capabilities.

Gardner (Gardner, 1993) developed a theory of multiple intelligences. He discerned certain fields in which students may have different aptitudes. These multiple intelligences give important information about students and the way they can learn. The number of intelligences varies a bit. Gardner started out with eight, two more have been formulated recently.

In Table 4.3, multiple intelligences are described (Heacox, 2002).

Table 4.3: Multiple intelligences.

Intelligence	Characteristics	Types of assignments
Visual–spatial intelligence	Good at reading maps, puzzles, enjoys drawing, painting, recognize patterns	Illustrate, design a brochure, design a post card
Linguistic–verbal intelligence	Enjoys reading and writing, is able to explain things well, good at debating	Debate, write a poem, make a folder, design a poster
Logical–mathematical intelligence	Good at abstract ideas, enjoys solving complex problems	Design a matrix, design a maze, analyze
Bodily–kinesthetic intelligence	Good at dancing and sports, good physical coordination	Play a role, do a pantomime, choreograph a dance, do a sketch
Musical intelligence	Enjoys music, both listening and playing, recognizes patterns and rhythms easily	Make a rap, write a song, make a jingle
Interpersonal intelligence	Good at interacting with people, skilled in nonverbal communication, positive relationships	Interview, help solve conflicts, work cooperatively

(continued)

Table 4.3 (continued)

Intelligence	Characteristics	Types of assignments
Intrapersonal intelligence	Good at self-analysis, good at analyzing theories	Keep a log, set goals, sum up ideas, reflect
Naturalistic intelligence	Good at characterizing nature, recognizes plants and birds easily, special attention for the natural environment	Scaffold ideas, solve a problem, design a walk, design a collection

The other two intelligences that have been formulated are the existential intelligence, focusing on the ability to think philosophically, being able to see the big picture. The other one that has been considered is moral intelligence, the ability to reflect and adhere to moral issues.

There are several tests that can be taken online to determine your aptitude in multiple intelligences (https://www.literacynet.org/mi/intro/index.html_).

Relating to the aptitudes of your students, will help them learn better. You can use these intelligences, for example, in designing tasks for students (Heacox, 2002). In Table 4.3, some examples of such assignments are given.

Assignment 4.1

Many tests are available online to determine the size of a short-term memory.

To get some idea you can try a simple test, first suggested by Johnstone (1997).

Give your students a date, say, for example, August 5.

Tell them to change that into numbers, that is, 8–5 or 5–8 depending on the way you write your dates. Then let your students arrange the numbers in the order from low to high.

The result would be 5;8.

They have to do this without any help, in their mind.

Then make it more complicated as shown below:

October 7 leads to 0;1;7

December 23 yields 1;2;2;3

March 8, 2019 yields 1;3;8;9

April 25, 1997 yields 2;4;5;7;9

Winding up with something like the following:

November 17, 1895 yields 1;1;1;1;7;8;9

You will notice most students drop off after five digits.

Assignment 4.2

Determine the sensory preference of fellow students.

Make groups of three. Two people should speak to each other, the third one should take notes.

Start a discussion about what you did last weekend/vacation. Person 1 can ask and keep the discussion going. Person 2 can tell from his or her recollection.

A visual person will use verbs and words like see, point of view, perspective, show, vision, shine, dark and light.

An auditory person will use words like loud, sounds, ask, harmony, speechless, noise, voices, hear.

A kinesthetic person will use words like bearable, emotional, feel, grasp, keep, move, tension and enjoy.

By marking the use of verbs in which the recollection is described, the preference may be determined in a few minutes.

5 Classroom management

Classroom management is a challenge; an important challenge. You need to be able to create a safe learning atmosphere for your students. The starting point for a safe atmosphere is mutual respect. The students respect you as a teacher, you respect the students and the students respect each other. This atmosphere is essential for learning to occur. It is only within an atmosphere like this that you are able to implement different learning activities.

You create a safe learning environment by interacting with your students. Both body language as well as verbal messages play a role. The challenge is that you need to find out what the effect of your actions is on students and their behavior. In some groups, classroom management is easy, in other classes it can be quite an effort. There are two ways of reacting to behavior in the classroom, and you should demonstrate both. You can give positive feedback on the behavior you wish for, and you can give negative feedback on the behavior you do not want.

One of the first issues that you need to understand classroom management is to exert your influence on a group of students or use your authority in reaction to behavior, has nothing to do with the way you normally interact with people. It is just one of those tasks you need to manage as a teacher. The more effective you are in getting the atmosphere you want in a classroom the easier it is to realize the teaching you want to do.

That students will challenge your authority, may be taken as a basic principle. The way they do that is subtle. They subtly try out your thresholds. They will try to find out at what point you will react one way or another.

A clear set of rules in your classroom will help you in establishing a classroom atmosphere that is relaxed, including a clear set of actions and activities that students will recognize. In the chemistry classroom and the chemistry laboratory, this is even more important as safety is an issue.

You can actually practice the behavior you would like to achieve. An example is the one-meter voice. When students are working in groups, you will notice that the volume keeps going up as the background noise is making conversation more and more difficult. A one-meter voice is a voice that is audible over a distance of one meter, but not much farther. Explaining this, and then asking students to use the one-meter voice will actually help. After 5 or 10 min, you can stop work and discuss with the students whether or not this works. When the noise level rises again, you can remind them of the one-meter voice.

Though giving positive feedback usually is not a problem, giving negative feedback and exerting authority is often problematic. One of the most important aspects in classroom management is your body language. The way you stand in front of the class should demonstrate that you are the central person in the room, determining what is happening in the classroom. Your bearing should radiate self-confidence.

https://doi.org/10.1515/9783110569629-005

Shoulders forward and head down beams out insecurity, while standing straight, shoulders backward and head up emanates self-confidence. Taking a class in play-acting will help you use your body in an effective manner, irrespective of how you really feel. Some people are very nervous, when they are in front of a classroom. Your mindset in front of the class is crucial, especially after you have had some issues with a group of students. You need to go in there with the idea that you are not going to let a group of teenagers sit on you. Once you have that mindset, it will show in your behavior in front of the class

Assignment 5.1

You need a number of steps often called the escalation ladder in your reaction to behavior.

There are escalation ladders for all sorts of situations, for love, war, conflict, stress and dog behavior, all indicate several levels of actions relating to a specific situation.

Students' behavior toward a teacher can be characterized at different levels:
- Questioning authority
- Arguing
- Noncompliance
- Defiance
- Verbal abuse
- Physical abuse

Your own behavior should prevent students from going to the next higher level. Students expect you to be authoritative, and know they are expected to yield to that authority. When a conflict arises, it should not get out of hand and escalate. Your action is supposed to be focused on dampening the conflict. One of the things you should never do is lose your temper. You can use an angry tone, or impatient sound, but you need to stay in control of your own behavior at all times.

A teacher's behavior to correct unwished behavior can be characterized as follows:
- Body language
  - Eye contact
  - Frowning
  - Shaking your head
  - Hand signal
  - Walking toward someone
- Verbal messages
  - Calling name
  - Calling name, and identifying wished action:
    - Please stop talking

- Please listen
- Please stop doing this/that
  - Second warning
  - Asking to come "see me after class"
- Physical action
  - Seating elsewhere
  - Sending outside the room for 5 min
  - Sending to principal

Assignment 5.2

Walking toward someone is actually very effective. Shortening the distance is intimidating, and will give your intervention more effect. If you do give a warning, you have to be certain, that you will fulfill the warning: "If you do this once more," or "If you bother me once more" "then …" You have to come up with the "then…" otherwise it will have a negative effect.

It is much easier to say: "I've warned you twice now, please come see me after class," leaving the action open.

If you do ask someone to see you after the class, it should be made clear you do not tolerate this type of behavior, and some sort of corrective measure should be enforced. A corrective measure should be uncomfortable for the student. A student should not get away with bad behavior. Having them work out the squares of the numbers 485 through 495 on paper often is a good deterrent.

Sending to the principal is a last resort to postpone solving the conflict at a later moment. You should solve the conflict with the student, with help of the principal and yourself, so he or she can return to your classroom without problems. It is extremely important that a conflict is solved before a student returns to your classroom. There should not be any lingering sentiments of ill will between a student and a teacher. On the other hand, it should also be clear what type of behavior will not be tolerated in class.

Having a clear set of rules of what is and what is not allowed in the classroom helps in establishing a good atmosphere. In the chemistry classroom this is vital, because of the safety issues involved. In Figure 5.1, the continuous line indicates the behavior you want in class. When students cross that line, they break one of your rules. Students need to become aware of these lines.

Assignment 5.3

Students will be trying to establish these boundaries for themselves. Your actions will define the picket line of wished for behavior. It takes time and practice to

Threshold reaction
before desired behavior

Threshold reaction
at desired behavior

Resulting behavior,
when reacting at desired
behavior

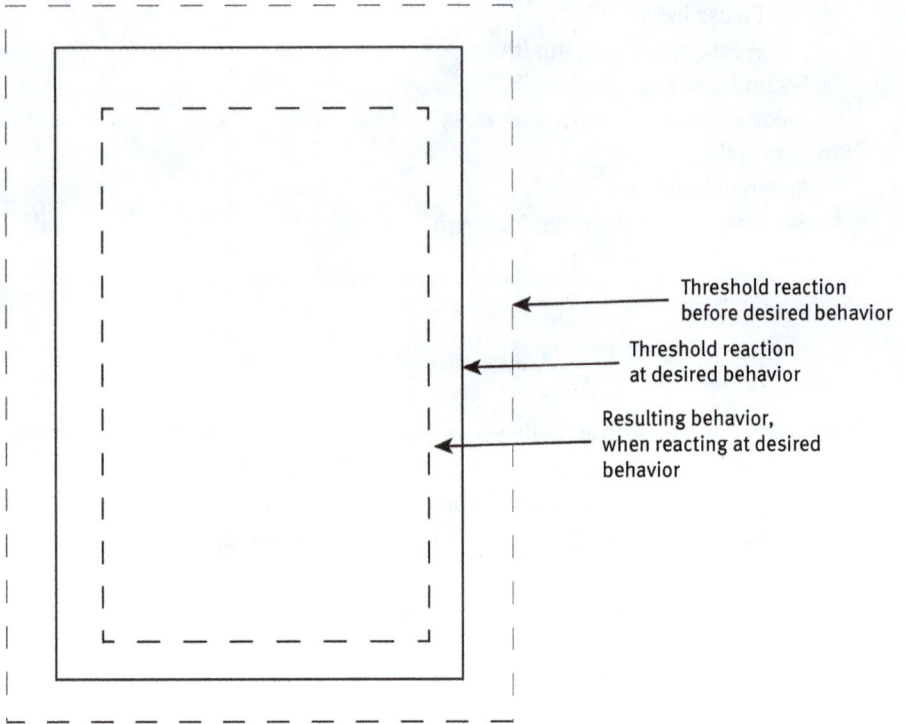

Figure 5.1: Reaction to behavior at different moments.

establish a set of signals that will work for you. The reaction to student behavior should indicate the boundaries of the behavior you find acceptable. You have to react to students' behavior before that boundary is crossed. Only in that case will you be able to attain the desired situation in your classroom, indicated by the continuous line. If you react to late, you will end up with a situation you do not want. One of the most important issues you need to react to is a show of disrespect. It may be showing disrespect to you or to another student. In both cases, reaction should be immediate and severe.

Students generally find a seating in the classroom based on their own social network. They sit together with the students they like, and are friends with. For learning, this is very often not the best seating scheme. When you start out with a class you can let them sit in a place of their choosing. After you get to know the students better, you should have the freedom to move them around the classroom. Not just to avoid social interaction between students, but also for teaching and learning practices. If you want to use cooperative learning techniques (see Chapter 7), you will need to combine the groups yourself. If you want to differentiate between the levels of students, you need to form three groups of slow, average and quick learners, who are seated together, so that you can address them as groups.

It is not easy to manage the behavior of 25–30 teenagers. One of the things you need to consider is that behavior can also have a background like boredom and frustration if a task is too easy or too difficult. Another factor may be a low self-esteem, or self-efficacy.

When you prepare your lessons in such a way that students are interested and motivated, you will have less and less problems with classroom management.

Assignment 5.1

Videotape, or have someone video tape you at the start of a lesson.

Comment on your pose. Are you confident or uncertain?

Show the tape to your fellow students and discuss how you might radiate authority.

Assignment 5.2

Formulate an escalation ladder for yourself, including a set of corrective measures. Corrective measures may include cleaning the classroom, removing the chewing gum from beneath the tables and so on.

Discuss these with your fellow students and add them to your portfolio.

Assignment 5.3

Formulate a set of rules for your classroom. There should be no more than 10.

Present the rules to your fellow students and come up with a final set of rules. And add these to your portfolio.

6 Use of the laboratory

6.1 Introduction

Chemistry is a branch of science where research in the laboratory plays an important role. In most natural sciences this is the case, but chemical research normally takes place only in a laboratory. Therefore, it seems logical to include lab work in the curriculum of secondary school chemistry. Practical work offers a rich environment for learning. Some of the learning goals that can be achieved with practical work have been formulated by Hofstein (Hofstein & Lunetta, 1982):

– *to arouse and maintain interest, attitude, satisfaction, open mindedness and curiosity in science;*
– *to develop creative thinking and problem-solving ability;*
– *to promote aspects of scientific thinking and the scientific method (e.g., formulating hypotheses and making assumptions);*
– *to develop conceptual understanding and intellectual ability and*
– *to develop practical abilities (e.g., designing and executing investigations, observations, recording data and analyzing and interpreting results).*

One of limiting factors for practical work inside schools are the facilities. Most schools have one lab with maybe one hood. To design and organize practical work within the framework of a secondary school is not always easy. The time spent in the lab by the students is another factor. If a lab period is planned in one lesson period, students do not have much time to perform all tasks. They need to plan carefully.

In a number of countries, a teaching assistant is available who can help prepare a lab session. He or she can make solutions and can have the equipment and glassware needed ready for the student. If you have to do that yourself, it takes quite a bit of extra time. The lab assistant will also clean up after the students.

6.2 Safety in the chemistry classroom and laboratory

As a chemistry teacher you are responsible for the safety in the chemistry laboratory. This includes the storage of chemicals and equipment. You need to look up the regulations that are applicable in that particular situation. Rules vary from country to country. Some rules apply to ventilation of the room, the type of floor to be used, requirements about the lab table surface and the number of exits (at least two).

The room needs to be large enough so that students can easily move, paths should not be blocked by chairs or stools and cupboards and tables should be stable. The floor should not become slippery. There are special requirements for water faucets and sinks as well as waste water disposal. The same goes for electrical outlets, as

https://doi.org/10.1515/9783110569629-006

well as gas taps. The room needs a central circuit breaker as well as a central cutoff for water and gas. There should be enough light in the room. The ventilation in the room should be such that the air in the room is refreshed three times an hour. These are only some of the requirements that can be applied to a laboratory.

The regulations for a chemistry laboratory are more severe than that for a physics or biology lab, due to the type of equipment used also the type of experiments being carried out is different.

Other regulations are about the safety equipment. You absolutely need a fire blanket, eye wash station, fire extinguisher and shower (see Figure 6.1–6.4) in case an accident happens. Particularly accidents with eyes will occur at times, because of carelessness. A poster indicating safety procedures and escape route should be pasted near the doors.

Figure 6.1: Fire blanket.

In the laboratory a copy of Material Safety Datasheets of the compounds stored should be available (see Figure 6.5).

Students should be aware that working in the laboratory requires a special attitude. Chemicals can bite. A set of specific rules should be printed and pasted at the entrance of the laboratory.

Figure 6.2: Eyewash station.

Figure 6.3: Fire extinguishers.

For personal safety, while working in the laboratory, you as well as the students should wear safety goggles and a white lab coat. Gloves as shown in Figure 6.6 are not normally used in the school laboratory. In the university

Figure 6.4: Shower.

labs, they have become standard by now. Long hair should be tied upand open shoes are not sensible.

The specific rules should include the following:
– keep pathways clear/ no bags in the room on the floor;
– long hair should be tied up;
– always wear lab coat and goggles;
– no experiments except those approved by the teacher;
– do not lift chemicals to eye level;
– never smell chemicals directly and
– no eating and drinking in the lab, including chewing gum.

In Figure 6.7 a more or less standard floorplan for a lab room is given

There are many possible varieties, but usually you will have four tables, with 6–8 students at each table, a large cupboard with stock glassware, a table for dispensing glassware, equipment and chemicals. Sometimes there is a special table for the teacher. There will be sinks, gas taps and electricity outlets at each table.

Things are all fairly straightforward if you are used to working in a lab. For students it is often the first time they work in this situation, and so they need to be taught how to behave.

One of the ways to impose the importance of safety in the lab, is by giving students a role within the safety regulation. You can appoint a student who will be

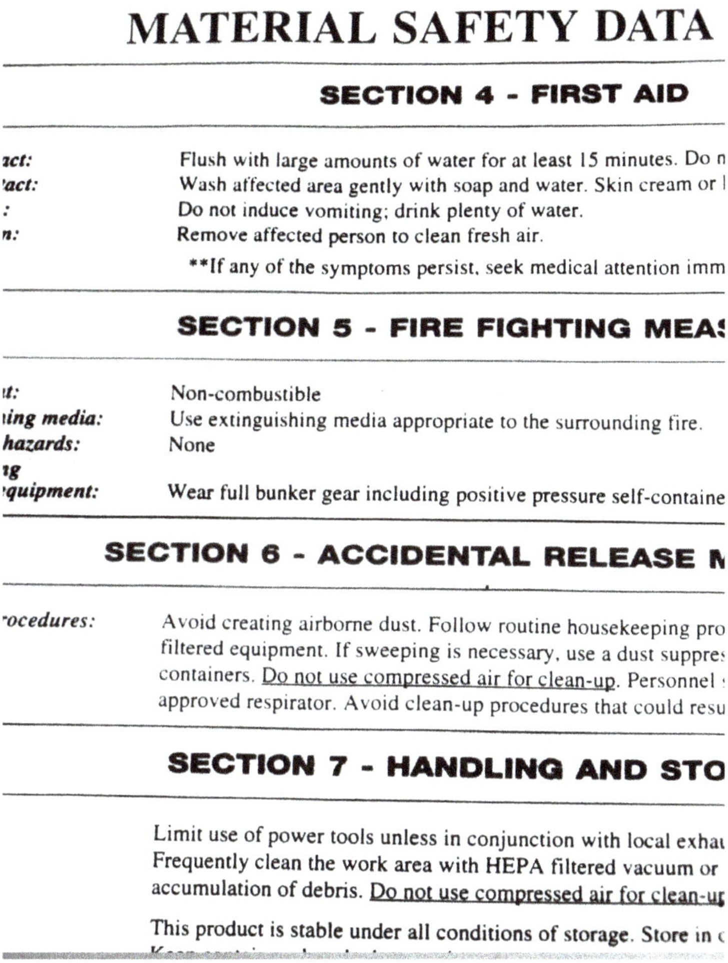

MATERIAL SAFETY DATA

SECTION 4 - FIRST AID

act: Flush with large amounts of water for at least 15 minutes. Do n
'act: Wash affected area gently with soap and water. Skin cream or I
:   Do not induce vomiting; drink plenty of water.
n: Remove affected person to clean fresh air.

   **If any of the symptoms persist, seek medical attention imm

SECTION 5 - FIRE FIGHTING MEA!

tt: Non-combustible
ing media: Use extinguishing media appropriate to the surrounding fire.
hazards: None
1g
quipment: Wear full bunker gear including positive pressure self-containe

SECTION 6 - ACCIDENTAL RELEASE N

ocedures: Avoid creating airborne dust. Follow routine housekeeping pro
               filtered equipment. If sweeping is necessary, use a dust suppres
               containers. Do not use compressed air for clean-up. Personnel :
               approved respirator. Avoid clean-up procedures that could resu

SECTION 7 - HANDLING AND STO

Limit use of power tools unless in conjunction with local exhat
Frequently clean the work area with HEPA filtered vacuum or
accumulation of debris. Do not use compressed air for clean-ur

This product is stable under all conditions of storage. Store in c

Figure 6.5: Part of a Material Safety datasheet in US-format.

responsible for safety at each table where students work. You can institute fire bud-
dies, where you make students responsible for each other.

Safety drills are important as well. Clearing the room in case of an emergency
should be practiced with each class at least once a year. A safety drill should include the situation in which you yourself are incapacitated.

Assignment 6.1

When using solutions of chemicals, it is best to use diluted solutions of maximal 0.1 M
concentration. Especially care should be taken with acids and bases of a higher
concentration.

Figure 6.6: Students with lab coat and goggles.

For safety issues, the amount of a chemical is one of the factors to consider. In general substances with a MAC (Maximal Accepted Concentration) value of less than 1 ppm should not be used in the lab.

Assignment 6.2

Microscale sets make it easier to perform certain experiments with your students. You will need much smaller amounts of chemicals, making it easier to comply with the safety regulations.

For organic chemistry, microscale sets are available that can easily be used in a school laboratory. Figure 6.8 shows the drawing of a setup for steam distillation of rosemary oil and Figure 6.9 shows the photograph of the actual setup.

Using microscale sets has advantages and disadvantages. For organic chemistry, it is ideal, as it gives you a chance to do some experiments in the classroom, linking theory to practice. For experiments in inorganic chemistry, the sets are different. In some cases, they have advantages, and in other cases normal scale experiments could be preferred.

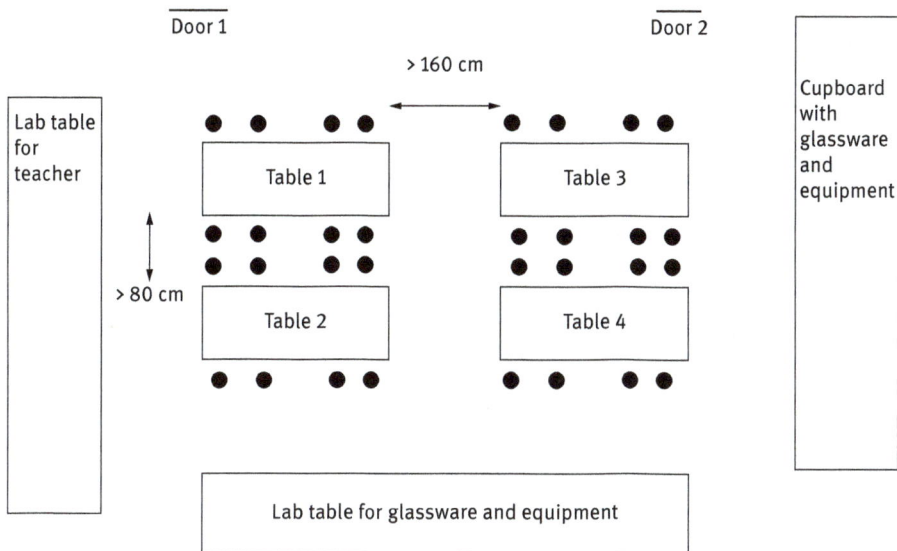

Figure 6.7: Floorplan for a student lab.

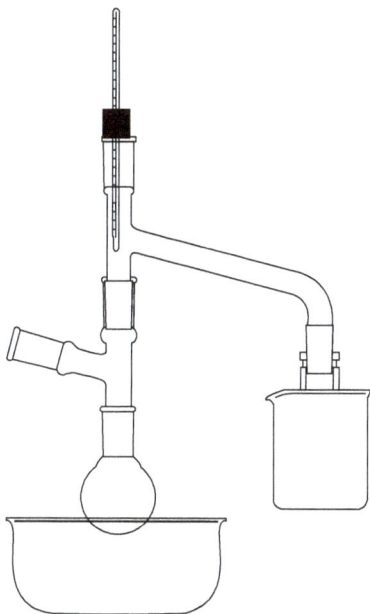

Figure 6.8: Schematic drawing of the setup for a steam distillation, using microscale glassware. From (J. H. Apotheker, 2005).

For your students to perform simple experiments in the lab, they will need a basic set of glassware. For special experiments, you can provide them extra equipment.

Figure 6.9: Steam distillation of rosemary oil using microscale equipment.

Such a basic set should include the following:
– test tubes
– a test tube rack
– a test tube holder
– a test tube cleaning brush
– a glass rod
– 100 mL beaker
– 250 mL beaker
– 100 mL Erlenmeyer flask
– 250 mL Erlenmeyer flask
– a funnel
– a spatula.

Waste management is the last issue in the safety procedures within a laboratory. Chemical waste calls for special measures of disposal. You will need special containers for different kinds of liquid waste. These are normally situated in a hood. For solid waste, you will need special containers for chemicals, broken glassware and refuse. There are special regulations in force about the way you can dispose off the specific waste. Normally, the easiest option is to cooperate with your local university for the waste management.

A specific safety risk is the use of a Teclu or Bunsen burner, depicted in Figure 6.10. The use of an open flame has quite some risks, especially with long hair not tied up. The use of a Bunsen burnershould belearnt properly. Students should be careful when leaving a burner on because the flame is often barely visible. Whenever possible, electric heating is preferred to heating on an open flame. In a university lab, open flames are hardly ever used. In organic chemistry experiments, the use is out of question because of the flammability of most compounds used.

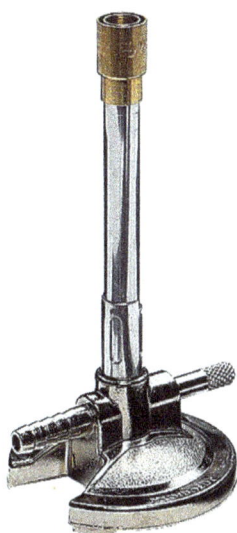

Figure 6.10: Bunsen burner.

6.3 Goals of practical work in secondary schools

Practical work in schools can have different goals. It can be focused on topics such as
- demonstrating a relationship between theory and practice
  - confirming theory
  - raising questions, arouse curiosity about theory
- developing practical skills
- promote aspects of scientific thinking and methodology
  - formulate research questions
  - design and execute experiments
  - observe

- record data
- analyze data
- interpret results
- communication about the experiment

It is important to realize that students enjoy doing practical hands on work. They enjoy the challenge; the extra precautions are challenging and exciting. In a number of cases, the results of an experiment are unexpected, for example, the burning of magnesium. First of all, the metal does not burn normally, and second the extremely intense white light gives a spectacular effect.

It is important to realize what are the learning goals of the experiments that students do in the course of a year. The training they get in doing practical work should enable them to do more of the steps in the scientific process, pursuing steps of inquiry independently.

It also helps students understand and remember scientific concepts better.

It is important that the training of the steps in scientific thinking and methodology are taught explicitly (Kruit, Oostdam, van den Berg, & Schuitema, 2018). To be able to do that students need to develop enough practical skills and insight.

Students can develop practical skills and insight by getting training in the school laboratory. An example is the introduction of titration experiments. These are often introduced in secondary schools, even though titration is not used much anymore in research. Before students can design experiments to answer questions such as "is there a difference in acetic acid concentration between different brands of vinegar?" they need to understand the nature of titration and they must have carried out several titrations to develop practical skills.

Assignment 6.3

You will find that most experiments are focused on establishing a relationship between theory and practice or on developing practical skills. Experiments and assignments focusing on developing inquiry skills are limited. In Table 6.1 lists the characterization of experiments, based on the level of inquiry (Buck, Bretz, & Towns, 2008).

Assignment 6.4

You will find that most experiments are either confirmation or structured inquiry. If you wish to do more, you will need to design experiments yourself.

One of the problems with most experiments is that they do not focus on inquiry, but more on handling physical objects than on reflecting and thinking about the

Table 6.1: Characterization of experiments based on the level of inquiry.

Characteristic	Level 0: Confirmation	Level ½: Structured inquiry	Level 1: Guided inquiry	Level 2: Open inquiry	Level 3: Authentic inquiry
Problem/question	Provided	Provided	Provided	Provided	Not provided
Theory/ background	Provided	Provided	Provided	Provided	Not provided
Procedures/ design	Provided	Provided	Provided	Not provided	Not provided
Result analysis	Provided	Provided	Not provided	Not provided	Not provided
Conclusions	Provided	Not provided	Not provided	Not provided	Not provided

More structure ◀━━━━━━━━━━━━━━━━━━━━━━━━━━━━━━━▶ Less structure

data. This means that there is little cognitive challenge in doing the experiments (Abrahams & Millar, 2008).

6.4 Planning practical work

Experimental work takes time and is expensive. The school needs to invest in building the laboratory. Maintaining a stock of glassware as well as chemicals and providing lab coats and goggles for the students are expensive as well. This demands a careful planning of experiments carried out by students. Both you and the students need to be well prepared. You will have to carefully formulate the learning goals that you have with the practical work, and the practical work should also be carefully evaluated.

When students carry out well-planned experiments, their appreciation is generally positive (Toplis, 2012). An important aspect is that students must have some sense of ownership of the experiments. This means that the students should be involved in some way in the planning of the experiment. This has to be done in the lessons prior to the lab session. When this is done, students will also be aware of the safety risks involved in the practical work. In most cases, it is not very difficult to adapt the cookbook type of experiments in most textbooks to more open experiments. Giving the students a role in the design of an experiment will greatly improve their commitment and appreciation.

Proper planning is essential, as the time available for the experiments is generally limited. This means that all solutions, chemicals and glassware need to be available at the beginning of the lab period. In addition, students need to be prepared. They have to be aware of what they have to do in the lab, so they can work efficiently. Before the lab period, the lab experiment has to be discussed. Students should be aware of the research questions related to the experiment. They should be able to formulate an answer to the question: "What is the goal of the experiment?" It will give direction to their actions and the learning process during the experiment. Special attention should be relegated to the safety during the experiment.

6.5 Planning inquiry-based experiments

When planning more open inquiry-based experiments, care must be taken to train the students explicitly in the different steps of the scientific procedure. Some of these steps are more complex than others.

6.5.1 Formulating research questions

To formulate research questions, students need to have some theoretical background. They need to know what is already known, what information there already available and what they understand about it before they are able to formulate questions.

Then they need to decide whether their question is a suitable research question, meaning it is a specific, single question, which can be answered by doing research.

6.5.2 Designing a procedure/experiment

To find an answer to the research question, they need to design a procedure or experiment. They need enough practical skills to assess what type of experiment is needed in order to find an answer to the question. To decide upon the proper procedure, they may need to perform a few experiments.

6.5.3 Observe and record data

Observing is not a straight forward process. One of the first things students need to learn is to separate observations from conclusions. If you let students observe the lighting of a candle and subsequently blowing out the candle, you will get different observations, while relevant observations may be missing. When observing and recording data, you are specifically looking at data relevant to the research question, while ignoring others. In the example of the candle, the color of the candle is not relevant for the burning process.

6.5.4 Analyzing data

Students need to learn how to make graphs and how to relate variables to each other. They need to be able to interpolate and extrapolate to analyze results.

6.5.5 Interpreting data

Relating the data back to the research question and existing theory is the main aspect while interpreting data. Comparing the results to other results from literature and discussing the accuracy of the measurements are all part of this step.

6.5.6 Conclusions

Formulating you conclusions, discussing the relevancy of the data gathered and formulating possible follow-up experiments complete the procedure. An evaluation of the whole procedure is fitting here.

6.5.7 Communication

Communicating about science has to be learned. Writing a decent report involving all the aforementioned steps takes time to learn. Communication about an experiment can be done in several ways, depending on the goal and target group of the communication. It is important that in communication about experiments, writing reports has more or less the same framework in all sciences taught in the school.

Assignment 6.5

Assignment 6.1

Draw up a safety protocol for the lab you will use in school. It should include possible messages to the caretaker in the school, contact with supervisors and fire brigade.

The safety protocol should include an evacuation plan of the room.

The size should be no more than A4 and should also be clear for visitors to your lab.

Discuss the plan with the people involved and hold a drill to see if it works.

Add the safety plan to your portfolio.

Assignment 6.2

Iodine (I_2) has a MAC value of 0.1 ppm Calculate the amount of Iodine (in grams) needed to reach that value in a lab room. On the basis of the answer, would it be allowed to let students carry out the sublimation of iodine in the lab?

Assignment 6.3 Learning goals of practical work

For the first grade of higher secondary education, go through the experiments for students in the textbook you use. Categorize them according to the learning goals:
- demonstrating the relationship between theory and practice
- confirming theory
- raising questions, arouse curiosity about theory
- developing practical skills
- promote aspects of scientific thinking and methodology

Find out whether there is a learning line between these experiments, do they become more involved and which skills are developed during the practical work. Compare your results with those of other students.

Assignment 6.4 Level of inquiry in practical work

For the next grade in higher secondary education, go through the experiments for students in the textbook you use and categorize them to the level of inquiry.

Which aspects of inquiry are missing in the experiments? Would it be possible to include these aspects in the experiments?

Compare your results with those of your fellow students

Assignment 6.5

Find out what frameworks/templates are used for reports in biology and physics. Relate the framework/template you wish to use to the frameworks/templates used in the other subjects.

Choose a framework, and compare with that of your fellow students. Add the chosen framework to your portfolio.

7 Assessment

7.1 Introduction

Assessment is a term that comprises a wide range of activities. It is more than marking on a paper or writing a test. There are several types of assessment used in education. Formative assessment (Vogelzang & Admiraal, 2017) is used by students and teachers to determine how far they have progressed in their learning. It is an important instrument for both teachers and students to determine whether they understood what they needed to learn. It also helps determine further learning steps in the classroomand does not play a role in grading a student.

Diagnostic assessment is used to determine the level of knowledge of students (Özalp & Kahveci, 2015). It can help determine possible misconceptions that students have and also determine or diagnose the level of expertise students have in a specific subject. It is an important instrument at the beginning of a course or a chapter. For both teachers and students, it gives insight into what still needs to be learned and which misconceptions need to be worked at.

Summative assessment is an assessment that is used to grade the students. It is used to determine whether or not they achieved the learning goals set at the beginning of a particular chapter. It also rates the students among each other, that is, it is used to differentiate an average student from a good and excellent student.

The above-introduced assessments are all teacher designed and are under the responsibility of them. There is another type of summative assessment, which is normally designed by a national agency. These national assessments are used to determine the level of knowledge and skills of students like the GCSE (General Certificate of Secondary Education, UK) (Earle & Davies, 2014). Quite some countries have national examinations at the end of a secondary school to determine whether or the students have attained some minimal level of knowledge and skills and also to rate them according to their relative ability. Finally, these examinations function as a benchmark for the level of education a school provides. In other cases, national exams are used as a condition for admission to higher education.

The Program for International Student Assessment and Trends in International Mathematics and Science Studies (TIMSS) (Le Hebel, Montpied, Tiberghien, & Fontanieu, 2017) are a form of international assessment, in which a sample of students in each country is tested. The results of these students are then used to rate the level of (science) education in the participating countries. These tests give some indication of the level of education in a particular country.

https://doi.org/10.1515/9783110569629-007

7.2 Criteria for assessment

In a leaflet published by the "Assessment Reform Group," based in Cambridge, UK, 10 principles about assessment are formulated (Broadfoot et al., 2012):

Assessment for learning
1. *is part of effective planning of teaching and learning;*
2. *focuses on how students learn;*
3. *is central to classroom practice;*
4. *is a key professional skill for teachers;*
5. *is sensitive and constructive because assessment has an emotional impact;*
6. *fosters learner motivation;*
7. *promotes commitment to learning goals and a shared understanding of assessment criteria;*
8. *helps learners know how to improve;*
9. *encourages and develops the learners' capacity for self-assessment; and*
10. *recognizes the full range of achievements of all learners.*

These principles reflect on all types of assessment. They describe the impact assessment has on the learning process of students. Assessment is part of the learning environment you create in your classroom, especially the results on summative assessment can have quite an emotional impact on students. These results determine the progress within the school and reflect on the students' position. The results can play a major role in the motivation of students and their commitment to learning goals.

The principles are based on five key factors for learning through assessment (Broadfoot, Gardner, Daugherty, & Gipps, 1999) that had been formulated before:
- *The provision of effective feedback to pupils*
- *The active involvement of pupils in their own learning*
- *Adjusting teaching to take account of the results of assessment*
- *A recognition of the profound influence assessment has on the motivation and self-esteem of pupils, both of which are crucial influences on learning*
- *The need for pupils to be able to assess themselves and understand how to improve*

When designing and using an assessment, these factors and principles should be taken into account.

7.3 The position of assessment in the learning process

In the whole process of teaching and learning, a major role is played by three related activities and ideas, which need to be aligned to each other. This idea of constructive alignment (Biggs, 1996) is depicted in Figure 7.1.

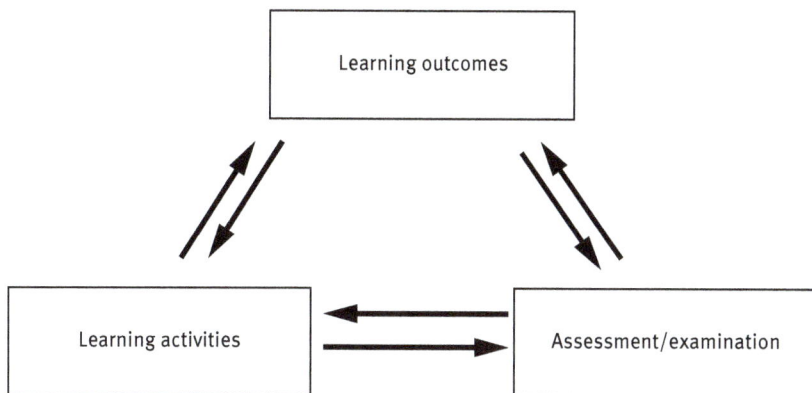

Figure 7.1: Constructive alignment according to Biggs (1996).

All three aspects are of equal importance. The learning outcomes determine the assessment and also the learning activities. The learning activities are linked not only to assessment but also determine in part the learning outcomes. When an outside assessment is considered, this influences the learning goals as well as the learning activities.

Students need to be prepared for a final examination in such a way that they can actually succeed in obtaining a score that relates to their abilities.

As a teacher you are involved in all three aspects. Although you are restricted in a number of ways, especially in the choice of learning goals, generally there is a national chemistry curriculum, or there are science standards you want to achieve. You will be using a textbook that has its own set of standards. In a textbook, choices will be made about which concepts and to which level students should learn them. The one thing you can depend on is that when you follow a textbook you will have covered the concepts that are in the national curriculum.

Depending on the type of course you are teaching you have more or less leeway in deciding which concepts you want your students to learn about. In many cases, you will want your students to attain a level of scientific literacy.

For all concepts you should decide for yourself to what level you will teach these concepts. Before you make this type of decision, it is important to consider the following questions (Loughran, Berry, & Mulhall, 2006):
– Why is it important that students learn about this concept?
– What do students need to learn about this concept?
– Which aspects of this concept need not be learned at this time?

Especially the last question is an important one, because it helps you determine more exactly what you expect your students to learn about a specific concept, in terms your colleagues and your students can understand.

7.4 Formulating learning goals

You need to be very specific about what you want your students to learn. Formulating your learning goals or expected learning outcomes should be focused on what your students should be able to do with the concepts you want them to learn about. Your learning objectives should be formulated in an active form which is often called SMART:

S: Specific, which means the objective should relate to an aspect

M: Measurable, which means it should be possible to assess whether or not the objective was reached

A: Attainable, that is, students should be able to attain the objective

R: Relevant, that is, the objective should be relevant to the subject discussed

T: Timely, that is, the objective should be achieved within the given timeframe

An example of a non-SMART-formulated learning goal:

A student should know the first 10 alkanes.

In SMART formulation:

- A student is able to give the names of the first 10 alkanes
- A student is able to write the structural formula of the first 10 alkanes
- A student is able to give the name of alkanes, which forms the backbone of an organic compound with no more than 10 carbon atoms.

The above are examples of more concrete formulated learning goals. Based on the first 10 alkanes, more learning goals can be formulated of course, for example:

- A student is able to give the names and structural formulas of isomers of alkanes with no more than 10 carbon atoms

In the SMART-formulated learning goals, there is a difference in the level of knowledge. The first two are mere reproduction. In the third and fourth examples, the student is asked to apply his knowledge in a new situation.

The level of knowledge is important as it is related to the attainability of a learning goal. The level of knowledge is an important issue in education. In order to determine the effectiveness of educational activities, it is important to have some measure of the level of knowledge a student has achieved. In the taxonomy of Bloom (Bloom, 1984), levels of knowledge have been described in six levels, which are indicated in Table 7.1. Bloom did not only look at cognitive aspects, which he called the cognitive domain, but also at the psychomotoric domain, indicating practical skills (Table 7.2), as well as the affective domain, indicating how well students identified with a subject (Table 7.3).

The three domains can be described in more detail, categorizing into general objectives and behavioral objectives, describing students' actions related to a specific subject. They describe the development of a novice to an expert in a certain field.

Table 7.1: Levels in the cognitive domain (Bloom, 1984).

Level	Description
Knowing	Recognizing and remembering facts, terms and basic concepts
Comprehending	Being able to organize, compare, interpret and describe concepts and ideas
Applying	Using acquired knowledge, facts and techniques
Analyzing	Breaking down information into component parts and determine how parts relate to each other
Evaluating	Make judgment based on criteria and standards, and reflect
Creating	Put elements together to form a new coherent whole, and reorganize in a new structure

Table 7.2: Levels in the psychomotoric domain.

Level	Description
Simple	Basic level of readiness to receive instruction and start activity
Imitation	Imitating and copying movement that is demonstrated
Manipulating	The development phase, where students are developing skills through practice. They are able to identify an action through a description.
Precision	Minimal errors in performing individual activities
Articulation	The coordination and performance of a series of individual activities
Naturalization	Automized performance of skills learned

Table 7.3: Levels in the affective domain related to science.

Level	Description
Receiving	Aware that science exists, and pay attention in class
Responding	Responds to the information, is willing to work with that information and enjoys the science
Valuing	Accepts scientific values, prefers scientific values and commits to values
Organizing	Builds up consistent value system toward science
Characterizing	Has integrated values in behavior

The cognitive domain is characterized by knowledge. Because of that it has had most attention, because student's progress has often been measured in terms of knowledge. It starts with simple knowledge and advances into more and more complex levels. The more cross-links and references are made, the higher the level of knowledge.

The affective domain has been neglected in education for some time, which led to severe problems (Rocard et al., 2007; Sjøberg & Schreiner, 2010) with students appreciation of chemistry and science. The affective domain deals with emotions, which are difficult to define, let alone measured in an assessment. In a number of cases, the attitude of students toward a course and the appreciation of a course can be measured through questionnaires.

The psychomotoric domain regards physical manipulation of objects such as glassware and laboratory equipment. Diagnosing of what can be improved in a certain setup, for example, is a higher level in the psychomotoric domain.

When you think of dissecting, for example, it means recognizing patterns and structures in the object being dissected. This takes time to learn. The same is true for microscopic work.

Learning to handle the microscope is one, but finding and identifying specific items is also part of this domain.

Several of the verb lists describing actions in a specific domain can be found on the Internet, using Google. Some examples are given in Table 7.4.

Table 7.4: Verbs relating to Bloom's taxonomy levels.

Cognitive domain		Affective domain		Psychomotoric domain	
Level	Verbs	Level	Verbs	Level	Verbs
Knowing	Define State Recall list	Receiving	Attend Ask Differentiate Accept	Simple	Observe Analyze Accept
Comprehending	Describe Interpret Compare Relate	Responding	Comply Volunteer Engage in Recite	Imitation	Copy Execute Follow
Applying	Solve Discover Select	Valuing	Support Argue Initiative	Manipulation	Re-create Perform Implement
Analyzing	Infer Contrast Discriminate	Organizing	Adhere Defend Explain	Precision	Demonstrate Show Calibrate
Evaluating	Reframe Devise Organize	Characterizing	Act Propose Revise	Articulation	Combine Coordinate Modify
Creating	Construct Develop Rewrite			Naturalization	Design Manage Invent

These verbs and related verbs can be used to define learning objectives at different levels. It is not always easy to differentiate between two levels. Hence, these levels are often adapted to certain extent. Another factor is that the number of levels is high. It is not always needed to differentiate between levels, hence, often the number of levels is reduced to four.

In further research, the group of Bloom combined some aspects of knowledge in the cognitive domain, in order to include actions that are related to knowledge (Lorin W Anderson & Krathwohl, 2001). The cognitive domain is divided into two different dimensions, each looking at different aspects of knowledge. They called it the knowledge dimension and the cognitive process dimension. In both directions, different levels can be discerned:

The knowledge dimension:
– Factual is the basic element students must know in order to understand a discipline.
– Conceptual is the interrelationships among these basic elements, combining into a larger structure.
– Procedural shows how to do things, methods to work within the structure, using skills, methods of research and so on.
– Metacognitive shows awareness about the way the related knowledge may be acquired, interpreting research results.

The cognitive process dimension:
– Remember to retrieve knowledge from the long-term memory
– Understand and construct meaning from instruction
– Apply, carry out procedures in a given situation and solve problems
– Analyze, break up a problem to its constituent parts, analyze the structure and apply the right methodology
– Evaluate and use results to draw conclusions
– Create and put different elements together to form a coherent system

Understanding these different levels is important in formulating your learning objectives, as well as designing the summative assessment.

Once you have formulated your learning goals as SMART-formulated objectives, you can start relating your learning activities and assessment to these goals.

7.5 Formative assessment

7.5.1 Introduction

Formative assessment is an important tool in the learning process. It helps students achieve the learning goals (Black & Wiliam, 1998; Vogelzang &

Admiraal, 2017). The formative assessment has to be directly related to the learning goals; otherwise, it will have no effect (Smith, 2007). The formative assessment activities need to be focused on helping the learning process of the students. Formative assessment should show the student what he knows and what he doesn't know. The results should have an effect on both the students and the teacher. For students, it should direct them to their weak points and stimulate them to work on those. Teachers should be able to give individual feedback on the learning process and help individual students improve. For the teacher, it will yield information on aspects that need to be discussed during class sessions, because some of the concepts have not been understood well enough. The main criteria for formative assessment are:

- it activates the students;
- it challenges the students;
- it demonstrates the students' proficiency;
- it guides the students in their learning process;
- can be done with individuals and in small groups;
- it gives information for both students and teacher about the learning progress.

It is actually better to let students work in small groups, because then they will start interacting and discussing about the problems posed. Students need to get feedback on the assessment in order to interpret the results they have achieved. The assessment needs to be evaluated in order to be effective.

7.5.2 Techniques of formative assessment

Homework
The most straightforward technique of formative assessment is exercises and solving problems. In exercises, students practice the relationship between variables in one-step calculations or questions. In problems, the number of steps involved in solving a problem and some sort of systematic problem-solving scheme are needed, as discussed in Chapter 9.

An example demonstrating the difference between an exercise and a problem is:

Exercise: What is the chemical amount of 3.78 g of sodium chloride in moles?
Problem: Calculate the resulting [Cl$^-$] in a solution made by dissolving 3.78 g of sodium chloride in 3.0 L of water.

Exercises and problems can be worked out in class rooms, or at home. One of the main advantages of working in the classroom is that you can let the students work in small groups on a problem. Together students are able to help and stimulate each other especially if you stimulate the cooperation. Within the realm of cooperative learning, many classroom activities have been designed to let students work together (Kagan, 1990).

Cooperative learning and assessment

Cooperative learning is more than just working together. When a cooperative learning group is formed, the group scores higher than a traditionally working group, or as an individual student (Figure 7.2).

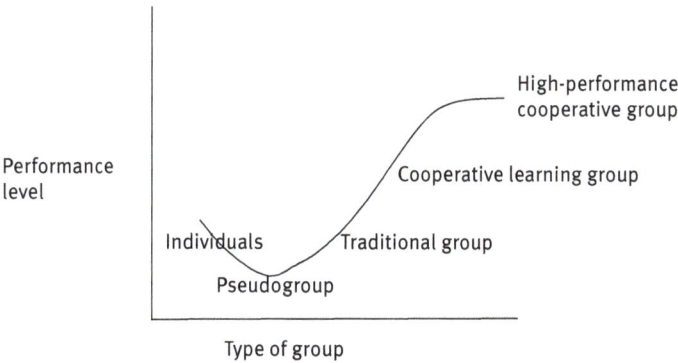

Figure 7.2: Performance level of groups (Johnson & Johnson 1999).

In order to let students work cooperatively a number of conditions must be met (Johnson & Johnson, 1989). The five conditions for cooperative learning are:
- **Positive interdependence**. Students need to have a dependent feeling on each other to achieve a goal, for instance, like a pitcher and catcher in baseball (the Red Sox won the world series on 29 October 2018). They need each other in order to succeed. If one fails the other one also fails.
- **Individual accountability/personal responsibility**. Students need to know they cannot let others do the work and just relax and watch the others do it. They must feel the personal need to learn and accomplish their part of the task.
- **Face-to-face interaction**. Students need to be able to see each other and interact directly with each other.
- **Interpersonal skills**. Students need to be able to interact in a positive manner with each other, giving each other positive feedback. Negative feedback, names calling is detrimental.
- **Group processing skills**. Students need to learn to work in a group, identify with the group and see the group as their unit.

Some of these conditions seem trivial, but not all are easy to achieve.

Face-to-face interaction is easily achieved by turning tables around, as shown in Figure 7.3.

Normal setup

Cooperative learning setup

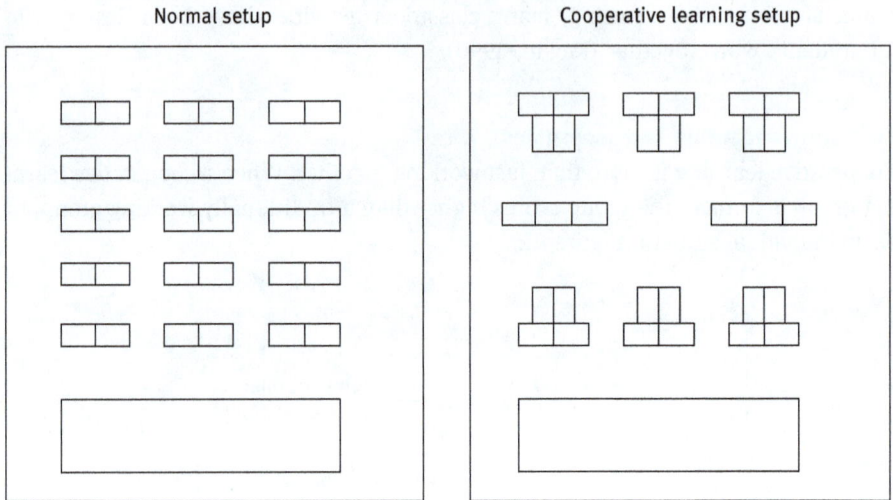

Figure 7.3: Switching a standard classroom to a cooperative learning classroom.

Students are not always used to work in group. Based on how they feel in the group, various groups can be formed. Table 7.5 describes some of these groups. Their performance level is indicated in Figure 7.5 (Johnson & Johnson 1999).

Table 7.5: Types of groups in cooperative learning.

Type of group	Description
Level 1: Pseudogroup	A group whose members have been assigned to work together but have no interest in doing so. This structure promotes competition at close quarters.
Level 2: Traditional learning group	A group whose members agree to work together but see little benefit in doing so. This structure promotes individualistic work with talking.
Level 3: Cooperative learning group	A group whose members work together to accomplish shared goals. Students perceive they can reach their learning goals if and only if the other group members also reach their goals.
Level 4: High-performance cooperative learning group	A group that meets all the criteria for being a cooperative group and outperforms all reasonable expectations, given its membership.

It is important to let the students move from a pseudo or traditional group to a cooperative learning group.

In a number of cases they do not really have the social skills to work together. Hence, to train them in working together, the so-called T-forms can

Listening to each other

Sounds like	looks like
Only one persons talks voice is only audible in the group	Others look at person talking others nod/smile

Figure 7.4: T-form.

be used (see Figure 7.4). A T-form is a form that looks like a T. It is often drawn on he blackboard. Together with your students, you discuss about how "listening to each other" sounds like and looks like. These properties are then written on the blackboard. Students then start working. After 5–10 min you ask them for their observations about this aspect of working together. In the discussion you can raise the questions, what went well and what could be done better.

These forms are very versatile and can be used to train any kind of behavior.

In order to meet the first two conditions of cooperative learning, you will need to formulate your assignments in a certain way. After a while students will get used to this mode of working, and almost automatically switch to a cooperative working mode. Before that is the case you need to induce them. There are very many classroom activities described for cooperative learning, which have a certain format (Kagan, 1990). When you use a format like that, these conditions are normally met. For working exercises or doing homework the activity which is based on "think-pair-share" can be used. The worksheet is given in Figure 7.5.

Because you use a worksheet, which can be handed in and checked by you, students tend to be more careful in writing completely worked solutions. Asking a second student to review the questions will be a learning exercise for the reviewer as well. The comment given should be more than just saying no, or wrong. By taking in the sheets you get an excellent idea about the progress of your students. By giving feedback on the comments and discussing these with the students, the comments will improve next time you use this worksheet.

Exercises §3.4, 16–19

Name student A:
Write the worked answers to the exercises in the box below

Pass the sheet to student B:
Name Student B:

Check pupil A's solution to question 16. In the box below write down whether you agree with his solution. Indicate why and if necessary give hints to better his solution. If something is not clear ask questions.

Pass the sheet back to student A
Student A. Read the comment B has written. You may improve your answer in the box above.
With a number between 1 and 10 indicate how sure you are about your final answer.

Certainty about the answer (1–10):

Figure 7.5: Worksheet used for "Think-Pair-Share" activity.

Another advantage is that you do the activity in the classroom, or as homework.

Students are dependent on each other for feedback. Because they have to work the solutions, they also have a personal accountability.

In another activity used for more involved problems students work together directly. This activity is called "Placemat." The problem is printed on a double-sized paper, about the size of a placemat (A3). Students each have part of the sheet to write on. They are not supposed to write on anything else.

Figure 7.6 shows an example that can be used during the introduction of organic chemistry.

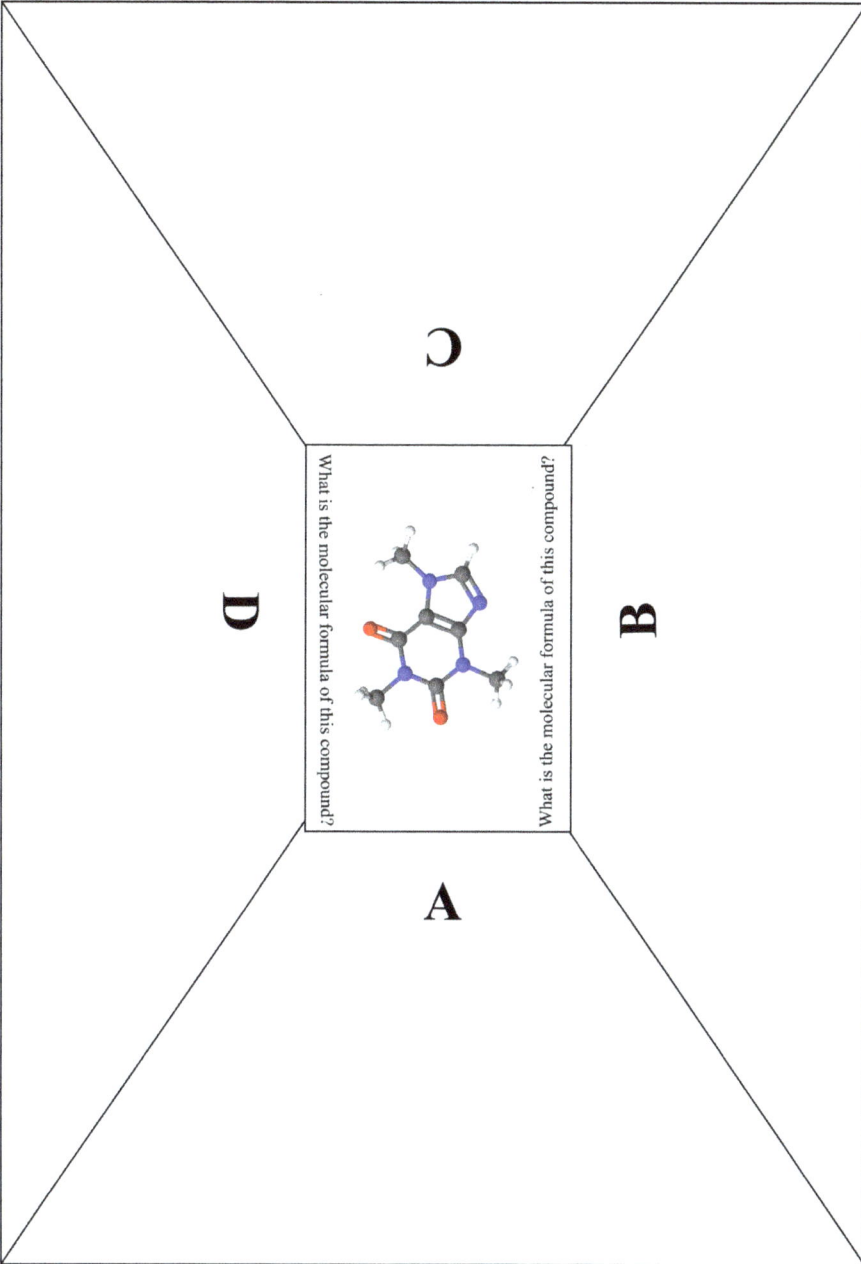

Figure 7.6: Example of placemat.

Assignment 7.1

Group discussion

Important in the learning process is that students discuss among each other what possible answers might be. A class discussion led by you is one of the ways in which you can assess what students know about a certain subject. Fellow students can react before you give an answer. It is to be preferred that students discuss among each other, before you take the lead again.

One of the ways you can start up a discussion by raising a question in a scheme is called peer-teaching (Mazur, 1997). An example of such a question is given in Figure 7.7.

Which of the following is not an isomer of hexane?

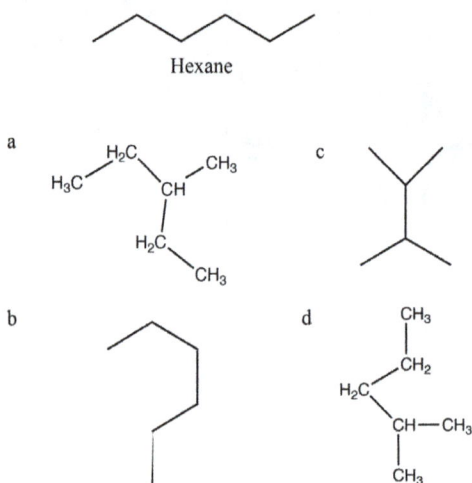

Figure 7.7: Example of a peer-teaching question.

The scheme is as follows:
- You pose the question
- You let the students first think about it in silence
- By showing of hands you tally their answers
- Then you let them explain to each other why they chose for a particular answer
- Finally, you tally their answers again by a show of hands
- Then you give the correct answer (b)

You'll find that it is difficult stopping the discussion, when they disagree among each other. Again, here you get an immediate idea of the skills of your students.

Assignment 7.2

Other activities

There is a myriad of activities that can be used to get students discussing and answering questions. Some of the examples are as follows:

- **Rolling a dice**. Each group of students gets a dice and a set of questions. The dice is rolled, and the corresponding question is answered.
- **Twitter board**. On a post-it students have to formulate in 140 characters what they learned during the lesson.
- **Questioning**. The students are asked to formulate a question on a sheet of paper about the concept discussed. The other students are supposed to answer the question. The questions are then circulated for answers among the other groups. After several rounds, the students receive their own question back. They rate the answers of the other groups.
- **Game playing**. When students are learning simple things like the symbols of the elements, bingo is a game that can be used. Students make their own bingo card using the symbols of the elements. The teacher reads the names of the elements.

Googling on formative assessment activities will yield many activities.

Assignment 7.3

7.6 Summative assessment

7.6.1 Introduction

There is a multitude of techniques for summative assessment. They vary from written tests, oral examinations, reports, essays, posters, multiple choice (MC) tests, line exams and so on.

Summative assessment means the assessment is used to score students. Summative can play a role in formative assessment as well. Formative learning is focused on the learning process, directing to more deep level learning (Bijsterbosch, van der Schee, & Kuiper, 2017). Care has to be taken on the effect of summative assessment, as it can also have a negative impact on students and let them focus on rote level learning. It depends very much on the type of test that is administered to the students.

One of the other dangers that can happen is the phenomenon of "teaching to the test". Especially in national examinations, teachers can focus much on preparing the students for the test, instead of focusing on broader learning goals. These national or benchmark tests are often different from the tests used in the normal classroom; hence, some preparation is needed, but a complete focus on these tests is detrimental for education.

Summative assessment is the instrument to determine at the end of a series of learning activities to what extent the students have reached the learning goals set at the beginning of the activities. Both for the teacher and students it is an important indicator, which will have an effect on motivation. This implies several criteria:

The assessment should be

– Valid
– Reliable
– Transparent
– Efficient

Valid means that there is a relation between the assessment and the learning objectives you have set in the course. It also means the assessment should be able to discern the difference in the level of students. High-performing students should score a higher grade than average students. Students that have not attained your learning objectives should not receive a passing score.

Reliability means that this particular assignment gives more or less the same results every time it is used. The results of this assessment should be comparable to earlier assessments.

Transparency means that the students are aware of what is required of them in order to receive a successful score on the assessment. They should know what they are expected to do, and which type of answers they are supposed to give, that is, there should be no ambiguity about the answers you expect from the students. For example, as an answer to the question: "where was the declaration of independence signed," a gifted student might answer: "at the bottom of the paper."

Transparency also means that students are aware of the assessment criteria, that is, they should be aware what is expected to score a high mark.

The effectiveness of the assessment relates to the way it is fast and efficient in reaching its goal. It also relates to the way an assessment contributes to the learning process of the student.

As mentioned earlier, a summative assessment not only assesses the cognitive domain but also assess the psychomotoric domain, and in some cases even the affective domain, as defined by Anderson and Krathwol (2001) and Bloom (1984).

In science, next to cognitive skills, practical skills also play a major role. These practical skills should also be assessed one way or another.

7.6.2 Design of a summative test

The knowledge dimensions mentioned earlier are an important part in summative assessment. The number of levels differs to some extent. Bloom uses six levels, and Krathwohl four. In international studies such as TIMSS (Mullis & Martin, 2015), three levels of knowledge are discerned:

- Knowing
- Applying
- Reasoning

In other situations (Miller & Kandl, 1991) and specifically in mathematics, education is often transcribed as:
- Knowing what
- Knowing how
- Knowing why

In some cases, these levels are subdivided into many levels (Bertona et al., 2014). Most often the apply level is divided into two levels:
- Apply in a known situation
- Apply in an unknown situation

That yields four levels of assessment, which can be viewed as a property of the assessment or of a quality of the student (Table 7.6).

Table 7.6: Levels of assessment.

Level	Property of the test	Quality student	Type of learning
1	Reproduction	Remember	Surface level
2	Apply in a known situation	Understand	Surface level
3	Apply in an unknown situation	Integrate with other knowledge	Deep level
4	Reasoning	Able to apply	Deep level

In the first level, reproduction is focused on the recall of factual information. MC questions are perfect for assessing this type of information. As a test form, MC exams are limited, as it focuses on recognizing terms and definitions. It is not a very active form of assessment. Examples are tests about the formulas used for elements, simple definitions, relation between names of compounds and their molecular formulas.

In the second level, students are asked to perform simple calculations they have practiced in class, or use some sort of procedure that was practiced earlier. For example, giving structural formulas of organic compounds, determining the formula of a salt and calculating the chemical amount of a certain mass of compound.

In the third level, the student is asked to apply procedural knowledge in a situation he/she is less familiar with. For example, with the given formula of salt, determine the charge of one of the components, and with the chemical amount and the mass of a compound, calculate the atomic mass of one of the components.

In the fourth level, the students have to determine how they can use the procedural knowledge in a new situation. They would need to analyze the problem in order to apply the knowledge they have. An example would be the following question:

Use freezing-point depression to calculate the molar mass, or the degree of dissociation of a given substance.

In order to design a test or to analyze a summative assessment, a test matrix is an instrument that is easy to use. It is primarily meant for written exams, but can also be applied in a broader sense. An example is given in Table 7.7.

Table 7.7: An example of a test matrix.

	Test matrix of						
Learning goal/ objective	Cognitive skills				Other skills		Number of points in a test
	Level 1: recall	Level 2: apply easy	Level 3: apply difficult	Level 4: reasoning	Affective	Psychomotoric	
1							
2							
3							
4							
Percentage in test							
Expected score of students							

This way you can link learning goals to questions/assignments in the test. You can also indicate the level at which you are testing your students. All levels should be present in a test.

The "percentage in test" gives you an indication of the level of difficulty of your test. The more level 3 and 4 questions you have, the more difficult is your test. For lower secondary, a distribution of
- Recall 30%
- Apply easy 30%
- Apply difficult 30%
- Analyze 10%

would be normal. In upper secondary, that is, in 12th grade
- Recall 10%
- Apply easy 30%

- Apply difficult 30%
- Analyze 30%

would be more expected. The percentages can be varied of course. In order to determine the expected score of a student you can set your expectations in the row "expected score of the students."

A fairly normal score would be that students score

- 100% on level 1,
- 75% on level 2,
- 50% on level 3 and
- 25% on level 4 questions.

This will yield a minimum number of points you expect your students to score. This number should have a relation to what you determine the passing score to be. Scores for tests are given differently in countries. The USA has a system of letters, A, B, C, D and F. In some European countries, 1 is the highest score, and 5 or 6 the lowest, Elsewhere the score is related to digits between 1 and 10 (highest). In most cases, a score of 50% is considered a passing score.

In order to determine the final score of a student, related to the points scored you will need to decide the cut-off point for a passing score. Figure 7.8 and 7.9 shows the scores of a group of students for a test that has a maximum of 72 points.

Scores in points

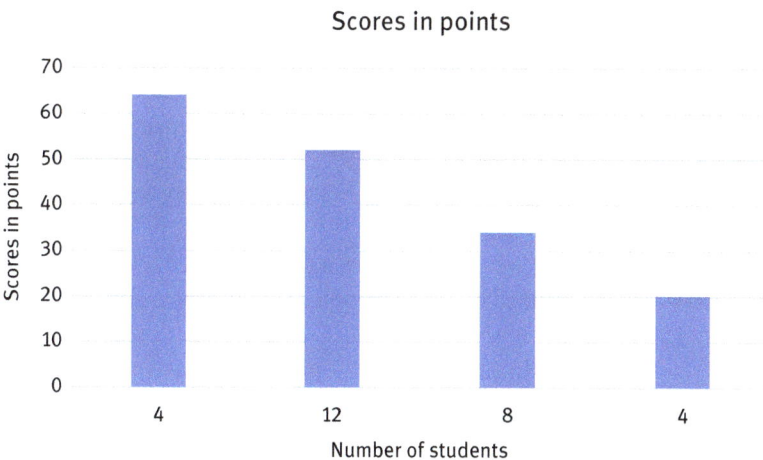

Figure 7.8: Average scores of a group of students (fictive).

This distribution of scores is a typical one for a class in secondary education. There will be some overlap, of course, as these scores have been grouped around an average. The group with an average of 64 points will be the group of excellent students, the group with an average of 52 will be the group of average to good

Number of students with a specific score

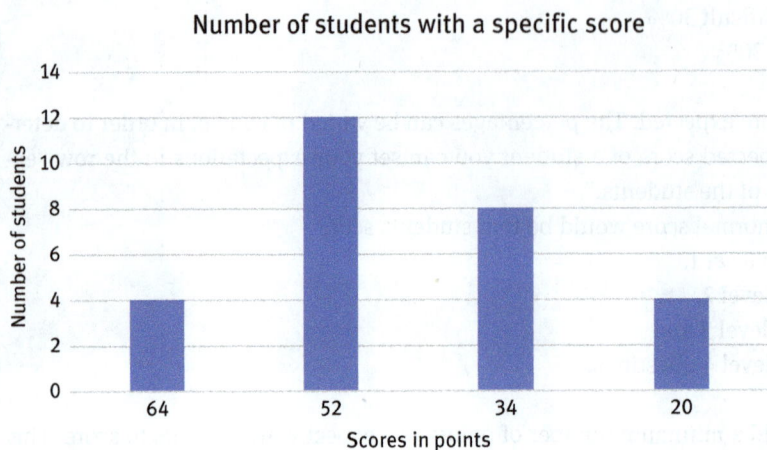

Figure 7.9: Distribution of average scores (fictive).

students and the group scoring around 34 will be the group of hardworking but not very talented students. The group scoring 20 has simply not prepared well enough for the test. Depending on your choice, some of the students in the group with an average of 34 will receive a passing score and some will receive a non-passing score.

Assignment 7.4

The length of a test is of course also an issue. In general, students have limited time to write a test. You can determine the time students need by asking a colleague to write the test. Students need at least two to three times as much time as a colleague would.

Nowadays, some students are labeled with terms like dyslexia and dyscalculia. They often need more time to read a test. For example, a test printed in double size will help them as well.

As mentioned earlier, the matrix is easiest to apply for a written test. The matrix will give you information about the reliability and the validity of a test. In order for the test to be transparent the student will need to practice the type of problems you want to test them with. The effectiveness of the test depends very much on the type of answers you expect from the students. The shorter the answers, the more concrete your assignments, the easier it will be to score the tests and to analyze them.

For science and chemistry, a written exam is one of the most used traditional ways to assess students. In chemistry, lab reports in varying degrees of detail are often used to grade students on practical work. Practical skills can be assessed by direct observation even though by now videos, the students make themselves are used as well.

There are of course other ways to assess students' skills and knowledge, writing a review of a number of articles, giving a presentation about a specific subject and making a poster about practical or theoretical research. Even making an exhibit for a science exhibition can be used (Apotheker, 2018).

When you are designing such a different type of exam the same matrix can be used to indicate the way learning goals are assessed, and the scores associated with the type of assessment.

Students will need precise instruction when you use this type of assessment. Writing a report of a lab experiment involves the use of several components that are not always clear for students. The way to formulate the goal of an experiment, or the way to report data needs practice.

Giving a presentation is something of a challenge even if PowerPoint, Keynote or Prezi are used. Students will need instruction in the use of these tools, but also in the way to prepare for a classroom presentation. On YouTube, several examples of instructions are given, like https://youtu.be/Iwpi1Lm6dFo and https://youtu.be/KbSPPFYxx3o about the use of PowerPoint and similar tools. More importantly, they need to know how to structure a presentation.

Making a poster or an exhibit for a science exhibition is even more a challenge. There are a number of steps that need to be taken care by the students while preparing for the exhibit and poster (Rocha de dos Reis, Marques, & Azinhaga, 2015) (Tables 7.8 and 7.9).

Table 7.8: Steps in research before building an exhibit.

Five basic research steps in building an exhibit		
Clarify research questions	Focus areas of interest	Students must thoroughly understand their questions before they can successfully conduct research – if questions are too broad, students may not know how to begin research.
Locate information	Use a variety of resources	Students should use multiple and diverse sources of information to answer their research questions – people, books, magazines and newspapers, videotapes, DVDs and CDs, Internet, sites, etc.
Summarize information	Take notes	Students need to summarize what they have learned – they will use their notes to answer the focus questions and develop their exhibits. To be able to effectively summarize, students must understand information at a fairly deep level and make decisions about what information to keep, to delete or to substitute.

(continued)

Table 7.8 (continued)

Five basic research steps in building an exhibit		
Analyze information	Examine notes to draw conclusions and answer research questions	Students are now ready to answer their research questions –developing a short written response to their research questions requires that they analyze their notes, prepare research conclusions and evaluate how well they have answered their questions.
Synthesize information	Share information with teammates to answer focus question and write a big idea and story line	Students are now ready to share findings with exhibit team members. This process forces students to step back from their independent research and integrate their collective knowledge. At this stage, students must adequately answer the focus question; in doing so, they will need to listen carefully to one another, synthesize information and evaluate the adequacy of their answers. When students are clear about their big idea, they are ready to answer the question "What do we want visitors to learn, feel and act in our exhibit?"

Table 7.9: Steps to build an exhibit.

Guiding questions for exhibit design		
What will we use to tell our story?	What objects will we select and/or build and what presentation methods will we use to display them?	Objects are central to exhibits as they are the visual devices that carry the story line, which include artifacts (real or created), models, interactive devices, video presentations, pictures, photographs, graphics, timelines, diagrams, charts and maps.
How will we get visitors to experience our story?	How will we make our exhibit relevant to visitors? How can we engage their senses?	Students must try to have in mind five principles: (1) relate to the visitor's personal experience; (2) reveal the big idea to the visitor; (3) use creative art forms to help tell the story; (4) encourage the visitor's curiosity, interest and questions; (5) present a whole story, rather than a part of a story. Displays that engage the senses are more likely to attract and hold the attention of visitors: students should consider ways to add visual, auditory, kinesthetic and tactile interest to their exhibits.

Table 7.9 (continued)

Guiding questions for exhibit design		
What will our complete exhibit look like?	What materials can we use to create our exhibit? How will we plan our space?	Creative displays can be made from ordinary supplies such as heavy cardboard, old shipping boxes, butcher paper, tension string and paint. Depending on where the exhibit will be housed, students may need a scale drawing for the entire floor plan. Developing this plan may involve the entire class, or the task could be assigned to a smaller group. The plan should specify the amount of space allotted for each display and the anticipated path of the visitor.
Will our exhibit work?	Will visitors like our exhibit? Will it be a cohesive whole?	Once students have designed their exhibits, they can conduct formative evaluation to improve their designs using their drawings. They can ask students from other exhibit teams, other students in the school, parents or other adults to respond to their exhibit ideas. Serrel and Raphling (1992) developed evaluation questions that might be useful during the formative evaluation process: (1) Do they like it? (2) Do they think it is fun? (3) Do they understand it? (4) Do they find it meaningful? (5) Does their understanding coincide with (or at least not contradict) the stated communication objectives for the element? (6) Does it give the user a sense of discovery, wonder or "wow"? Based on what students learn from the formative evaluation, they may want to brainstorm alternative, better ways to design their exhibits. Prototyping helps students test their assumptions with visitors before they go too far in the exhibit development and design process – before they are inclined to stick with their ideas and design, even if it doesn't work for visitors. Prototyping is actually an iterative conceptual design process. Students design the mock-up, talk to visitors and redesign based on visitor input.

An exhibit is ideal for scoring students not only on creativity but also on their understanding of the concepts they have worked on. This type of test is very motivating for students, as it gives them a specific role in a process. They feel responsibility and ownership of the exhibit.

An example of an exhibit given in Figure 7.10 was produced as part of a module about the difference between breastmilk and cow's milk (Apotheker & Teuling, 2017). This exhibit in science exhibition was viewed by about 3,500 people.

Figure 7.10: Example of an exhibit built about the difference between breastmilk and cow's milk.

Posters and wall papers/news sheets also motivate students. Producing a booklet in which their contributions are rendered is highly motivating. Posters can be used in a contest, which also stimulates students to work.

The main effect of these types of summative assessment is that they motivate students. It also leads to deep level learning, as they have to integrate the concepts and skills they learned in a concrete product.

Assignment 7.5

7.7 Grading summative assessment

7.7.1 Written exams

This will be the type of exam you grade the most. The first thing you need to do when you have designed a test is to make a marking scheme. You will assign points to each part of the test. These points are added together to give a possible maximum score. In general, you will then relate the scored points to a score on the test. As discussed earlier, the final score will depend on the type of scoring used in your situation.

Rating an answer to questions or problems can be done in two ways. You can either deduct points for mistakes made, or you can give points for correct answers. Consider the following question in Figure 7.11:

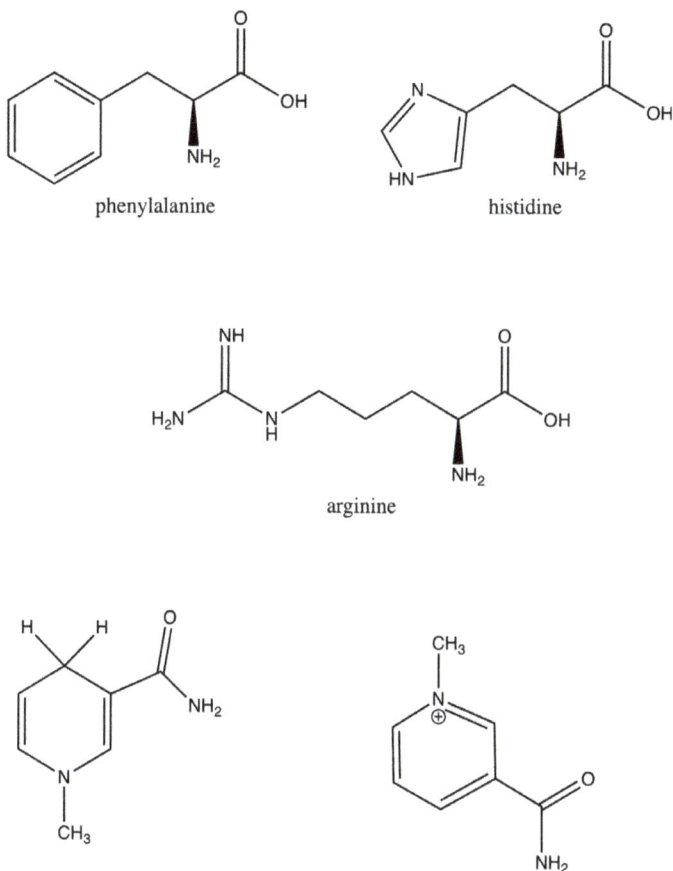

phenylalanine

histidine

arginine

Figure 7.11: Sample question in assessment.

In a correct answer, 1, 2 and 5 would be indicated. This could lead to 3 or 6 points, 1 or 2 for each correct answer. If only 1 or 2 are marked, a student would receive 2 or 4 points. How many points will you award if the fourth compound is also marked as being aromatic? What happens when a student marks all five? This last one is easy, no points. You need to consider the variables beforehand. A wrong compound would result in deducting 1 or 2 points.

This is a fairly easy type of reasoning. In more complicated questions (Figure 7.12), you will need to identify the steps needed in the calculation (this question was adapted from a central examination in 2012 in the Netherlands).

A detergent used in a dishwater may contain sodium percarbonate. Sodium percarbonate contains 1.5 mole H_2O_2 per mole carbonate. When sodium percarbonate is dissolved the hydrogen peroxide is released in the solution.

The volume of water in a dishwater is about 5.0 L.

Calculate the hydrogen peroxide concentration in dishwahing liquid when a tablet containing 2.2g sodium percarbonate is used.

Figure 7.12: Example question. Adapted from HAVO examen 2012.

Steps in the calculation are:
- Calculate the molar mass of sodium percarbonate (157.0 g)
- Calculate the number of moles of sodium percarbonate in 2.2 g
- Calculate the number of moles of hydrogen peroxide in that amount of sodium percarbonate
- Calculate the concentration of hydrogen peroxide
- Give the answer in two significant figures

Most logical would be awarding 5 or 10 points for the question. When mistakes are made, points may be deducted.

Assignment 7.6

When you start grading an exam you will want to be objective and score each student as objectively as possible. There are, however, some effects that can hinder your objectivity:
- **Signific effect** means there is the change of focus in marking. Some mistakes are emphasized more during marking. You get irritated because the students make the same (stupid) mistake.
- **Halo-effect** means you take into account the way an answer is written down, either sloppy or very neatly. Some students are very sloppy in the way they jot down answers. This may have a negative effect on the score, because you have to look for answers.

- **Sequential effect** is the effect that after marking a mediocre answer, a better answer is marked higher than it would otherwise be.
- **Contamination effect** means you take into account who it is that answers. If you know whose work you are grading, you tend to think, I know what he means, or you might think for him that is a very good answer.

Two other aspects that influence grading are related to the following effects:
- **Norm shift** means you adjust the norm after you have seen a number of similar mistakes and become more lenient.
- **Restriction of range** usually occurs with rating lab reports or other types of exams, which means not the full score range is applied but only scores are given between 4 and 8. Scores lower than 4 are not given, but also none higher than 8 (often with the argument that 9 is for the teacher and 10 is for God).

Once you know these effects may occur, there are some simple rules that help in avoiding these effects.
- Mark a test question by question. That helps in grading all students the same way.
- Make sure you score the problems anonymously. Change the sequence in which you score the test after each question.
- If you want to change the marking scheme for a specific question, make sure you apply the change for all students.

One of the other effects that should be considered is the "interrater reliability". This means that the score on a question should not be dependent on who is marking the exam. In quite a few cases one coordinated test is used for a whole year group of students with different teachers. The score teacher A gives for a test should be within 85–90% of the score that teacher B gives. It is obvious these differences should not occur. It is, however, seldom that this aspect is checked, while differences may be large.

Assignment 7.7

7.7.2 Grading other work

Grading things like essays, lab reports, posters, presentations, articles and exhibits is a challenge. They are different from tests, and have no clear-cut answers that are correct or incorrect. Creativity of students is involved, which may or may not be taken into account.

One of the first steps is to assign criteria to the product you want the students to make.

For a poster attention points may be:

- Title
    - Readability
    - Fits content
    - No spelling errors
- Content
    - Accurate
    - Consistent
    - No grammatical and spelling errors
- Labels
    - Consistent use of labels
- Graphics
    - Attractive use of graphics
    - Help in understanding content
    - Correct use of labels in graphs

In a lab report, other attention points will be used.

An easy way to grade this type of products is the use of rubrics. Examples are given in http://rubistar.4teachers.org/index.php. Rubrics are tables, in which the left column contains the criteria you wish to use. The other columns contain short descriptions of what you expect from these criteria. An example, created with the help of the website rubistar (n.d.) is given in Table 7.10.

Assignment 7.8

When you use a rubric for grading you will find to change the descriptions, in the columns related to scores. When you start assessing the work it will become clear what you would like to see in a product. Specifically, the difference in quality will become more evident. In general, you need to use a rubric at least once before it is complete.

Rubrics are a great help to improve "interrater reliability" when more than one person is grading the products.

Determining a final grade depends on the weight assigned to the criteria, as shown in Table 7.10, where each scientific concept and method counts for 40% of the final score.

Again, it is up to you to decide how you will determine the final score on a product.

Rubrics can help students as well. A rubric will indicate the conditions needed to get a passing score on a product.

Table 7.10: Rubric for a lab report created with the help of rubistar.

Criteria (% in final score)	Excellent (A)	Good (B)	Sufficient (C/D)	Not sufficient (F)
Components of the report (10%)	All elements are present and additional elements that were felt needed	All required elements were present	One element was missing	Several elements were missing
Diagrams/graphs (10%)	Clear graphs and diagrams are used when needed, and labeled neatly and correctly throughout	Requested graphs and diagrams were used, and labeled correctly	Graphs used were labeled incorrectly or some labeling was missing	Requested diagrams and graphs are missing and/or labeled incorrectly
Scientific concepts (40%)	Accurate and thorough understanding of concepts	Most concepts are understood correctly	Limited understanding of concepts	Concepts not or incompletely understood
Methods (40%)	Methods are described completely, including all details	Methods are described completely	Methods are described, some aspects are missing	Incomplete description of methods used
Other criteria				

7.8 Evaluation of a test

Suppose that you get the results in a test indicated in Figures 7.13, 7.14 and 7.15.

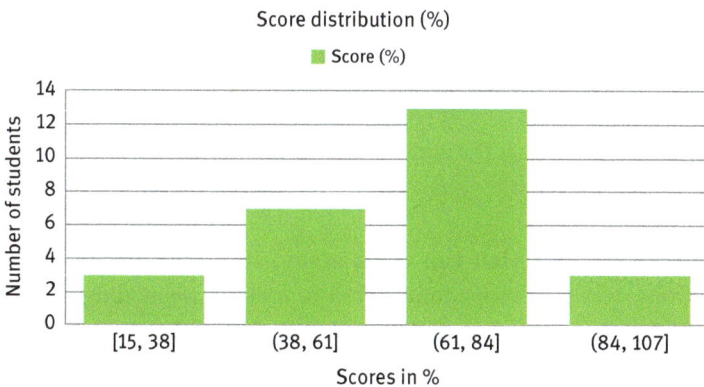

Figure 7.13: Distribution of scores in a normal test (fictive).

Score distribution (%)

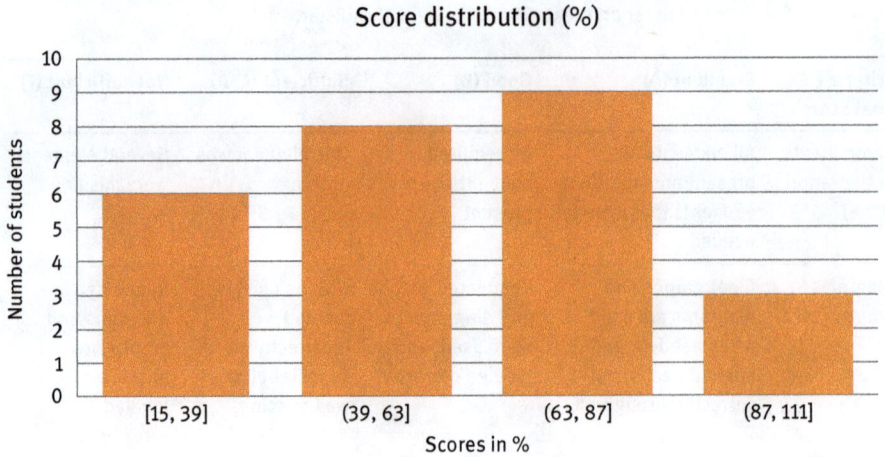

Figure 7.14: Distribution of scores in a difficult test (fictive).

Score distribution (%)

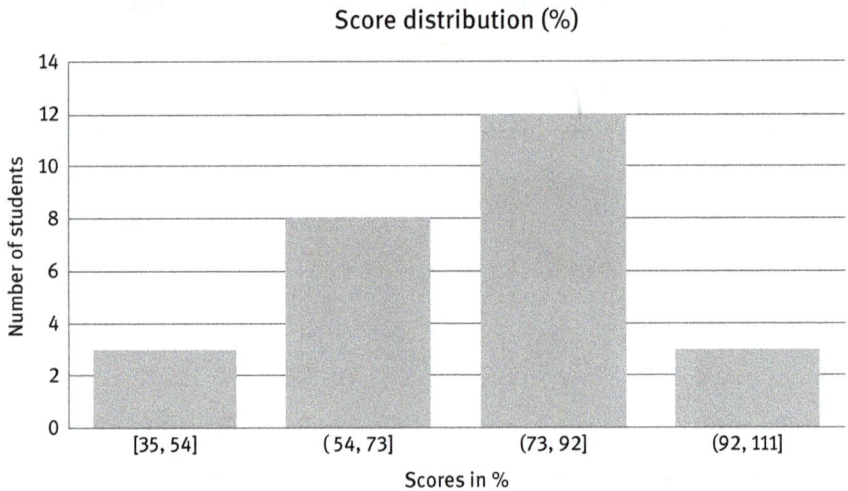

Figure 7.15: Distribution of scores in an easy test (fictive).

These are all fictive of course but can occur easily. They give an indication about the difficulty of a test that was administered to a class. Depending on the rules you may decide to change something in the way you grade the test.

When you evaluate a test, you need to look at the points scored on each item. When there are items that only one or two students scored correctly, which may be an indication that those items were too difficult. Sometimes you realize that there was a mistake in a test question. In that case, you will want to delete the question or at least not consider it in the final score.

The evaluation of summative assessment is an important tool that can be used to evaluate your teaching activities. It will tell you whether you reached your learning goals. It will give you information about changes you want to make in your teaching activities. It is important to do this directly after you have finished teaching a chapter.

Assignment 7.9

Assignment 7.1

Try out one or both of the worksheets in your own classroom.
   Report back to your fellow students or tutor about what happened.
   Add your report to your portfolio.

Assignment 7.2

Formulate a peer question for yourself and try it out on your students. The question should query one concept at a time and should be straightforward. One of the options is often

    d. not enough data to make a choice.

When students go for that option, or are reluctant to choose an answer, most likely the question is too difficult for them.

Assignment 7.3

Find an activity on Internet that you think suitable, adapt it to your own use and try it out in the classroom.

Discuss the activity with your fellow students or your coach.

Add the activity to your portfolio.

Assignment 7.4

Take one of the tests that was used by your coach in a previous year, and analyze it according to the matrix given in Table 7.7. Fill in the matrix and discuss the result with your coach.

    You can also use a test you have designed yourself.

    Add the test and the matrix to your portfolio.

Assignment 7.5

Design an alternative type of summative assessment involving the creativity of your students.

Make a test matrix linked to the assignment to indicate the learning goals you wish to assess.

Administer the test to your students.

Evaluate whether or not you were able to assess your learning goals.

Using a questionnaire find out how the students rate this type of assessment.

Present your findings to your fellow students/coach.

Add to your portfolio.

Assignment 7.6

Design a test for your students, using a test matrix and prepare a marking scheme.

Let your students write the test and grade the tests using the marking scheme.

Evaluate whether the marking scheme was adequate, and take into account all possible mistakes.

Add these documents to your portfolio.

Assignment 7.7

Copy a question from a test you have used and include the marking scheme. Select the answers from a student that scored full marks, a student with half the number of points and a student that received no credit for his answer. Copy these as well, without giving the marks you gave and without any marks.

Ask your fellow students (or teachers) to do the same.

Grade the questions you received.

Then compare and discuss the marking of each question.

Formulate ways how you might enhance the "interrater reliability."

Assignment 7.8

Design a rubric for grading a lab report or another more or less open assignment you will use in your classroom. Use it for grading the assignment.

Evaluate the use of the rubric.

Discuss the evaluation and the use of the rubric with your colleagues.

Add the rubric and evaluation to your portfolio.

Assignment 7.9

After grading the final test and after analyzing the results of your students, write a short evaluation about your teaching activities for the part that was covered by this test. Indicate whether you reached your learning goals. Discuss what went well, and what could have gone better. Indicate what you wish to change next time you teach this material.

Add the evaluation to your portfolio.

8 Inquiry-based learning

8.1 Introduction

In the beginning of a teaching career, you may let your colleagues and textbooks guide you in your teaching schedules. You need time to practice and experience to learn how to implement teaching activities in your lessons. You need to find out what you feel are effective classroom activities and which are not. That takes time. Once you feel comfortable in teaching, it is time to start challenging yourself again.

One of the challenges may be to start designing parts of your lessons. Doing a project or changing the order in which you introduce a specific concept are the things you will want to do after you have gone through the entire curriculum of the subject. You can try out things in small scale, just a few lessons, or you can try focusing on a whole chapter. Whatever you do, you (and your students) need to feel comfortable with the changes you make.

There are several models that can be used to make changes. Inquiry-based learning is one, context-oriented learning is another, problem-based learning is yet another term used. They all have a number of aspects in common. They activate students' mind. Students are much more involved in their own learning process, determining what knowledge they need to acquire.

There are pitfalls, however (Kirschner, Sweller, & Clark, 2006; Sweller, Kirschner, & Clark, 2007). You have to realize that you still need to guide your students in the new processes. Students are not expert researchers, and need to be shown what are the proper ways of gathering knowledge through reading literature as well as deciding which literature is valid. Formulating proper research questions, setting up experiments, interpreting results are the factors in which students need to be trained.

The important conclusion is that these types of project cannot be done with minimal guidance. You need to let your students take small steps in acquiring experience over time. If you do this, the results can be overwhelming. In Dutch secondary schools, students do a short thesis of 80 h in a subject of their own choosing. Several universities, as well as the Dutch Academy of Sciences (https://www.kna wonderwijsprijs.nl/), award prizes to the best thesis. The results are astounding; the level of research is very high. This can only be achieved by proper training.

The steps students go through in these models of teaching are all similar, with changes only in the details.

One of the main questions often asked is why you would implement this type of education. As indicated above (Kirschner et al., 2006), these projects have pitfalls. On the other hand, these projects have some influence on the students with an inclination toward science (King, 2012). Their learning does not improve much, but they are generally better able to use their knowledge in a new situation. They also

https://doi.org/10.1515/9783110569629-008

get a much better idea of science at the university level, making a possible career choice easier. This improved appreciation of science is an important factor (Rocard et al., 2007) for the future of science education.

One of the problems you may encounter when you start using these methods of teaching is opposition from colleagues and students. When everybody is used to a certain type of workflow, making changes can be difficult. When students are used to sitting back passively, listening to teachers and now have to be actively engaged, they may and will protest. The introduction of this type of teaching activity takes more preparation time, especially when it is new for teachers. This needed extra time may also be a large threshold. Another factor is that teachers have to take up a different role. The students are more on the center of the classroom activities compared to the teacher. The teacher has more the role of a coach or director, on the sideline, while others take action. That is often out of the comfort zone of teachers (Butcher, Brandt, Norgaard, Atterholt, & Salido, 2003).

8.2 Problem-based learning

In problem-based learning (PBL) a specific situation is taken as the central problem. For example, the question: *"If we exhale carbon dioxide and inhale oxygen, why would mouth-to-mouth resuscitation be effective for a person who is not breathing?"* (McPherson, 2018) can be the central problem in a PBL exercise, which in this case entailed two lessons of which one was in the lab.

The students then go through steps that may be guided (Wallace, Bernardelli, Molyneux, & Farrell, 2012):
- Gather information by asking questions like:
  - What information is available?
  - What questions can I ask?
- Formulating goals
  - What is the task?
  - What is the problem that needs to be solved?
  - What knowledge do I need?
- Brainstorming
  - Who can help me?
  - Where can I find out more?
- Action taking
  - What will I do?
  - What problems will I have to solve?
- Evaluation
  - What was the result of what I did?
  - Have I solved the problem?
  - What have I learned anything?

– Communication
    – How do I communicate my results?

The question asked in the beginning of this section was answered by determining the difference between the average molar mass of atmospheric air and the molar mass of exhaled air. This was done by emptying a syringe and filling it with atmospheric air or exhaled air. The ideal gas law was used to determine the average atomic weight of both types of air.

The amount of guidance provided and the time spent on different steps can vary. The learning goal of the teacher was to let the students practice with the ideal gas law. By using this PBL exercise, students gain a deeper understanding of the relationship between the different properties of a gas, and better understand how the ideal gas law may be used.

While assigning a problem-based learning activity, classroom activities based on cooperative learning are often used. In the chapter on formative assessment these are discussed in more detail. Classroom discussion with the students may help them find responses to the problems they need to solve.

8.3 Context-oriented learning

As discussed in an chapter 1, students did not appreciate chemistry very much. Since this has become clear, several methods have been introduced to change that attitude (Borley et al., 2016; Demuth, Parchmann, & Ralle, 2006; Eisenkraft & Freebury, 2003; Heikkinen, 2002). The common factor in these new approaches is that they link chemical concepts to everyday life.

Mahaffy (Mahaffy, 2004) introduced the idea of adding an extra dimension to Johnstone's triangle, which is shown in Figure 8.1.

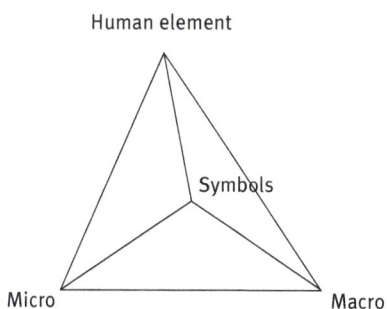

Figure 8.1: Johnstone's triangle with human element added.

His idea was that the human element should be used at any level to introduce chemical concepts. Later this idea was used to link contexts to Kansanen's didactic

Context

Teacher

Content   Student

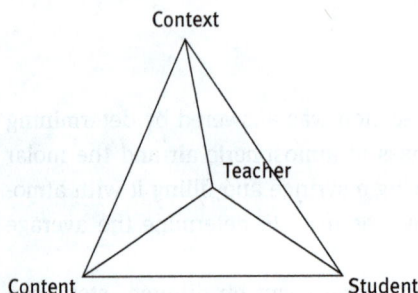

Figure 8.2: Didactic triangle linked with context.

triangle (Kansanen, 2003) as shown in Figure 8.2 (Apotheker, 2004). This idea clarifies the role a context can play in the interaction between teachers, students and chemical concepts.

More traditional education in chemistry has been aimed at discussing conceptual knowledge about chemistry. Vital and abstract concepts are used to interpret standard situations and solve standard problems. Practical work is mainly used to illustrate principles and practices (Tytler & Australian Council for Educational Research, 2007).

Contexts, if any, are introduced as an example of the use of concepts.

There are several definitions of what a context is. A common definition is:

Context, then, is essentially conceived in terms of a sociocultural setting, calling for tool-mediated actions, operations, and goals that are to be valued in the framework of that activity. (Oers, 1998)

Somewhat reworded, a context is an authentic (meaning it is not fictional) situation that can be studied from a chemical perspective, or a situation in which chemistry or chemical concepts play a major role. Such a context can play a role in engaging students.

The major focus of context-oriented chemistry education has been to attract students more toward studying chemistry and to relate chemistry more to the society (Apotheker, 2006; Nentwig, Demuth et al., 2007). The idea being that when students can relate chemical concepts to aspects of their daily life, they will be more interested in studying chemical concepts, relating the macroscopical world to the microscopical world of molecules and atoms.

There is a wide range of contexts that can be used. The difference between cow's milk and mother's milk, for example, has been the context for the study of biochemistry (Apotheker & Teuling, 2017). In order to explain the difference, students need to know the characteristics of fat, carbohydrates and proteins.

One of the first problems a teacher needs to solve is finding a context that matches the concepts that you want your students to learn. This is actually contrary to the original idea of context-oriented chemistry education (Butcher et al., 2003).

A context-based approach is when the "context" or "application of the chemistry to a real-world situation" is central to the teaching of the chemistry. In such a way, the chemical concepts are taught on a "need-to-know" basis; that is, when the students require the concepts to understand further the real-world application. (King, 2012)

The need-to-know basis places some problems on the chemical concepts. If, for example, the mouth is chosen as the context, the chemistry involved will be about the buffering systems in saliva, the enzymes in saliva, the structure of teeth, the use of polymers to fill cavities in teeth, the taste receptors in the tongue and so on. This involves a wide range of chemical concepts that are not normally related to each other in more traditional chemistry courses. This can lead to relinquish the use of contexts (Butcher et al., 2003).

In the development of a curriculum using contexts, the contexts are carefully chosen in order to fit the introduction of concepts as they would also be introduced in a more traditional course (Apotheker, 2009). Perfume, for example, is used to introduce organic compounds and to learn the IUPAC rules for naming organic compounds.

Looking back at the tetrahedron in Figure 8.2, you can discern different faces of the tetrahedron linked to different steps working in context-oriented education. The first face defines the link between the teacher, the content and a context. The teacher is active in finding a context linked to the content he wants to teach. This is also the preparation step in which the teacher designs the educational activities he wants the students take part in.

In the German project "Chemie im Kontext" (Nentwig, Parchmann, Gräsel, & Ralle, 2007) four distinct steps are set in the design of a lesson series. These are illustrated in Figure 8.3. There are more ways in which a lesson series can be designed.

The tetrahedron is then turned to face 2, in which the teacher interacts with the students and introduces about the context. In "Chemie in Kontext" this was phase 2. The students get an introduction about the context and formulate what they already know about it. Within 'ChiK" these questions were completely free, in later stages they became more guided by the teachers, focusing on the chemical concepts that are supposed to be discussed. In the module about breast milk, the first question students have to answer is, "Why don't babies just drink milk from the supermarket?" (see Figure 8.4).

They are then shown a table with the composition of milk of various mammals (Table 8.1).

The third face comes forward when the student starts working with the content related to the context. Students answer the questions they formulated in the previous phase; in ChiK this was called the elaboration phase.

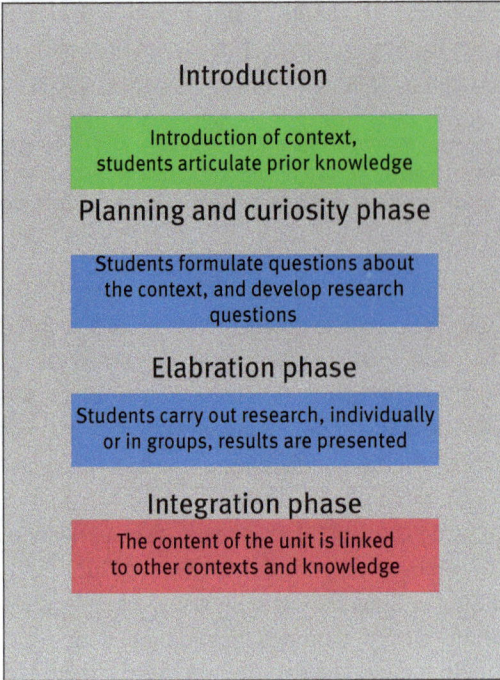

Figure 8.3: Steps in "Chemie im Kontext".

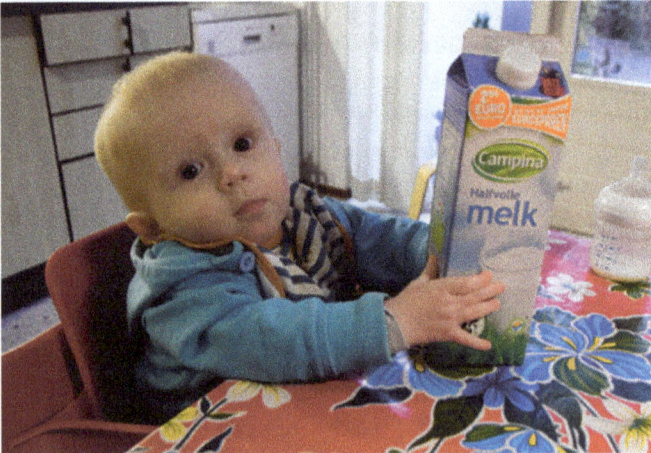

Figure 8.4: This is also milk, right?

Table 8.1: Composition of milk in various mammals.

	Human	Cow	Goat	Sheep	Horse	Buffalo	Donkey	Reindeer
Protein	1.5 g	3.5 g	3.8 g	5.2 g	2.1 g	4.0 g	NA	NA
Fat	4.0 g	3.4 g	4.1 g	6.2 g	1.3 g	8.0 g	1.4 g	18.0 g
Lactose	6.9 g	4.6 g	4.4 g	4.2 g	6.3 g	4.9 g	6.3 g	2.8 g
Residuals	0.3 g	0.8 g	1.9 g	0.9 g	0.4 g	NA	0.4 g	1.5 g

Source: Wikipedia: http://nl.wikipedia.org/wiki/Melk_(drank)

In the final face the teacher and student work together to link the knowledge of chemical content to other knowledge the students already have. The content is de-contextualized and integrated with the knowledge students already have. In ChiK, this was called the integration phase. These faces are shown in Figure 8.5.

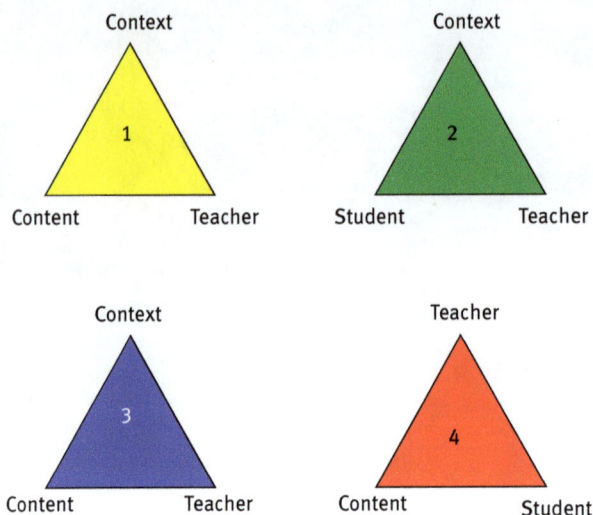

Figure 8.5: Four faces of the didactic triangle.

There are several ways in which you can design your educational activities using a context. The main idea of using a context is to demonstrate the students the role chemistry plays in the society. Everyday issues are linked to the chemistry they learn. This way the chemistry they learn is not directly connected to the concepts of chemistry they need to learn. They feel they need to learn chemistry in order to understand the societal phenomena they are studying. This is normally very motivating for the students.

There is a special group of contexts that may be used. These UN development goals ("UN sustainable development goals," n.d.) play an important role in many discussions about worldwide societal goals. These are shown in Figure 8.6.

In quite a few of these goals chemistry plays a major role, sometimes very obvious, as in the goal for clean water and sanitation, or zero hunger. Linking the chemistry you teach to a context taken from one of the UN development goals will link to the motivation of students to make a better world for themselves and their children.

8.4 Inquiry-based learning

Inquiry-based learning is a very broad term used to describe activities in which students are activated by involving them in some type of research in which they play a

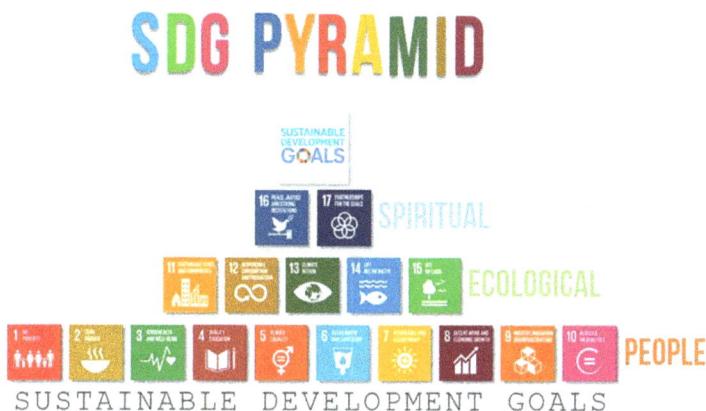

Figure 8.6: UN development goals as shown by Unity in Diversity (www.unitedindiversity. org). Reprinted with permission from United in Diversity Creative Campus @ Kura Kura Bali, Copyright 2016.

more or less active role. The main idea is to activate and stimulate students in order to motivate them, and to let them appreciate chemistry more (Hemraj-Benny & Beckford, 2014).

Inquiry-based science education (IBSE) has been introduced as a way to improve students' interest in science (Osborne & Dillon, 2008). There are several ways IBSE can be introduced (Gejda & LaRocco, 2006).

When studying the scientific process of inquiry, or rather the scientific method, several steps was formulated. These steps can be taken by the student or presented to them by the teacher. These steps are described in the first columns in Table 8.2 (Bybee, Powell, & Towbridge, 2007). In the remaining part of the table the possible student activities that can take place are formulated.

The list of activities is not complete, and could of course be formulated in another way. The idea behind the table is to demonstrate the difference in the role of the students. Student-directed learning allows the student to be active, while teacher-instructed learning leaves students with little freedom. One of the background ideas in inquiry-based learning is that it should be student directed. Students do not have the background to take all these steps by themselves. They need to learn how to do this. It is important for teachers to realize that as a teacher he/she has to train his/her students toward the steps mentioned in the left-hand side of the table. This can be done in many ways (Exploratorium, 2006). It is important that each step in the process needs to be executed separately. It also means the role of the teacher changes.

Table 8.2: Steps in inquiry.

Student-directed learning→	A lot			Little
Teacher-directed learning →	Little			A lot
Scientific process ↓				
Formulating research questions	Student formulates questions	Student selects questions/ formulates new ones	Student rephrases questions based on teachers or question given in books	Student uses questions provided by teacher or book
Gathering data relevant for research questions	Student determines which data are needed and collects them	Student gets instruction to gather certain data	Student gets data and is asked to analyze them	Student gets data and explanation about how to analyze them
Formulating explanations based on the data	Student analyzes the data and discusses the results	Student is guided in the process of discussing and analyzing the data	Student receives different ways to analyze the data and use them for discussion	Student receives analysis and conclusions from data
Connecting explanation to scientific knowledge	Student looks for literature connected to his research for conclusions and discussion	Student is directed to relevant literature	Student receives relevant literature	Student receives direct information about other literature
Communicating about and justify explanations	Student presents his/her findings in a logical and precise manner	Student is trained in communicating his/her results	Student gets general instruction about the way results are to be presented	Student gets precise information about the steps and guidelines for his communication
Reflecting on learning process	Student evaluates himself/herself based on feedbacks received	Student is trained in self-evaluation	Teacher discusses the evaluation process with the student	There is no evaluation

A template for the design of inquiry-based learning that has been used widely is the 5E method (Bybee et al., 2006). In this template five steps are described that the students need to go through in an inquiry-based learning activity.

In Table 8.3, these steps are described.

Table 8.3: The 5E model.

Steps	Description	Techniques used
Engage	Getting interested in the subject of the lesson series, acquire a sense of ownership with the subject, students want to learn more.	Students gather information, are asked questions about the subject, may do some introductory experiments.
Explore	Students pose central questions about what they want to learn and study in this module.	Students formulate what they need to know in order to understand the chemistry behind the context better.
Explain	Knowledge is gained, data are collected and scaffolded, teachers and the students scaffold the knowledge about the content.	Students gather data, connect to existing scientific knowledge.
Elaborate	The attention shifts to another subject, in which the knowledge can be used.	The knowledge gained will beis used and applied to a new setting
Evaluate	Students are tested on their knowledge of the content, the module is evaluated as well	Students take a written test for a summative assessment.

As an example to illustrate the steps, a module about perfumes can be used (Apotheker, 2009). In the engage phase, students read an article about the substances related to scent that play a role in perfume. In the explore phase, one of the questions that comes up is how you can distinguish the different molecules and how you can name them. They find a group of compounds that are similar, for example, esters. They learn some of the chemistry of esters, and synthesize aspirin as an example of an ester. They also learn about saponification.

In the elaborate phase biodiesel is introduced, where the students are expected to apply the knowledge they have acquired about esters in the production of biodiesel. They actually make some biodiesel in an experiment, when asked how to dispose off oil waste from the fat used for deep-frying in the school kitchen.

The 5E methods has been extended to include more steps (Eisenkraft, 2003; Apotheker, 2018).

In Table 8.4, the steps in the 5E model are compared to a more traditional teacher-centered type of teaching.

Table 8.4: The 5E method related to "normal" teaching.

Step	5E	Normal
Engage	Creates interest, stimulates curiosity, calls for questions, demonstrates current knowledge	Explains concepts, gives definitions and answers, gives conclusions
Explore	Allows students to cooperate, observes and listens to students, gives students time to reflect problems	Explains how problems can be solved, tells students what is wrong, gives information that will solve problems, leads students step by step to the solution
Explain	Stimulates students to find explanations, asks for proof built on experiences of the students	Introduces not directly related concepts and skills, does not ask students for explanations
Elaborate	Expects students to use definitions, encourages students to apply the learned concepts in a new situation	Gives answers, uses direct instruction, explains how the problem should be solved
Evaluate	Looks for clues if students understood the concepts, tests with open questions	Tests vocabulary, loose facts, uses multiple choice

The steps offer a convenient framework that can be used to design the different steps in an inquiry-based lesson series. The roles of the teacher and the students can be varied, as described in Table 8.2.

Assignment 8.1

Assignment 8.1 Using the 5E Framework

The 5E framework can be used for a series of lessons, but the steps can also be designed for use in one lesson.

Design an activity for your students in which you use the 5E framework, formulating each step carefully. Note at which stage listed in Table 8.2 you position the activities in your design.

Use the design in the classroom and evaluate the lesson (series).

Compare your series with those of your fellow students.

Add the design to your portfolio.

9 Chemistry in upper secondary education

9.1 Prerequisites

When a student takes chemistry as a subject in upper secondary school, one expects certain knowledge. Students must have studied science and chemistry in lower secondary. They should be able to explain the relationship between the molecular and atomic world and the macroscopic world. Processes and properties at the "micro" scale are related to the properties at the macroscopic level. Students should know the use of symbols and formulas in chemistry, representing either the "micro" or "macro" scale.

They should be able to describe the molecular model and Dalton's atomic model. Students should know that molecules are very small and that they are present in large number. On the basis of these models, students should be able to describe the difference between a pure substance and a mixture, as well as the difference between an element and a compound.

They should be familiar with the Periodic Table and should be able to indicate the position of the nonmetals in the Periodic Table and to give the symbols used for a number of elements. In addition, they should be able to use these symbols for recognizing and understanding the meaning of the chemical formula of a compound and to balance a chemical equation. They should understand and explain about a limiting reagent. They should be able to draw structural formulas of simple compounds, including the first four alkanes.

Students are expected to explain the role of chemistry in the preparation of drinking water. They should be able to indicate the difference between chemical energy and other forms of energy as well as to explain the difference between an exothermic and an endothermic reaction. They need to understand and describe the need for energy transition in the world.

The aforementioned knowledge will form a solid foundation to build upon while starting for upper secondary schools. It is important to check whether all students have achieved these learning goals before you start your lectures. This should be done in the first week of class. You will find that some students know more, and then there are those whose knowledge is not up to the mark or some have misconceptions about these subjects. Again, it is important to clear all the misconceptions.

https://doi.org/10.1515/9783110569629-009

Assignment 9.1

9.2 Mathematical and general science skills

In lower secondary school, students have a general introduction to science and mathematics. For upper secondary, a number of basic skills as well as knowledge and definitions of certain concepts are needed.

Basic skills are related to the scientific process and include the following:
- Formulate a simple research question
- Formulate a (simple) hypothesis
- Describe the difference between observing and interpreting
- Use measurement instruments
- Plot a graph in the correct way, using the correct labels
- Analyze results
- Interpret a graph
- Arithmetic skills
  - Calculations with fractions
- Algebraic skills
  - Solve equations with one variable
- Use formula's for quantities and properties
  - definitions of quantities and properties
  - Use units in the proper way
  - Use scientific notation
- Write a simple report

Most of these topics must have been discussed in lower secondary or upper primary classes. It is important to check whether your students have mastered these skills.

Lack of arithmetic and mathematical skills is one of the most common problems that occur. Most often the understanding of numbers is weak. This can be demonstrated by letting students use the scientific notations and work a number into a scientific notation:

$$3,478 = 3.478.10^3$$

In addition, the use of prefixes in properties and quantities often is a problematic area:

$$70 \text{ cm} = 0.70 \text{ m}$$

Using formulas, like the formula for density: $\rho = \frac{m}{V}$, is often an issue. Students have problems calculating the third property/quantity when the other two are given, especially while calculating the volume when mass and density are given.

Students often find dimension analysis difficult at this level. One of the ways to work on these skills is to give the students some exercises in each lesson so as to practice the skills needed to work with the chemical concepts discussed in class. To maximize the knowledge transfer between different subjects, it is best to use exercises that come from mathematics.

When discussing isotopes and isotopic compositions, one of the standard problems that students face is calculating the standard atomic weight based on the isotopic composition of an element. Conversely, students are asked to calculate the isotopic composition of an element from the standard atomic weight. It helps to go back to the mathematics lessons where these types of calculations were introduced and help the students practice with exercises not related to isotopes before attempting to solve the problems with isotopes, as decimal numbers are the additional difficulty in such cases.

Assignment 9.2

9.3 Topics in upper secondary chemistry

Before you start teaching, you should know the topics you need to cover in upper secondary chemistry classes. Most often there will be a curriculum. In any case, there will always be the way chemistry is taught, the text books that are used, and so on. An issue is the time available for teaching. In most countries, it is three years, that is, grades 10, 11 and 12 for teaching upper secondary. In some countries, students need to make a second choice, after the first year, as to at which level will they take chemistry. For example, in the International Baccalaureate, there is standard-level chemistry as well as high-level chemistry. This is the case in several countries.

Curricula are constantly undergo change. Every 5–10 years, some major changes occur, but the basic topics remain the same in most cases. A good list of topics to refer to is published by the International Chemistry Olympiad (IChO). Within the IChO, topics are defined under three categories, which are published every year in the so-called preparatory problems ("Regulations of the International Chemistry Olympiad," 2013).

- *Level 1*: These topics are included in the overwhelming majority of secondary school chemistry programs and need not be mentioned in the preparatory problems.
- *Level 2*: These topics are included in a substantial number of secondary school programs and maybe used without exemplification in the preparatory problems.
- *Level 3*: These topics are not included in a majority of secondary school programs and can only be used in the competition if examples are given in the preparatory problems.

The level 1 topics normally are the ones taught in advanced level courses. Level 2 and 3 include topics for first- and second-year university. In normal courses, in secondary education the level is slightly lower.

In general topics included in most secondary school curricula are as follows:
- Atomic models
    - Rutherford's model of a nucleus with protons and neutrons and an electron cloud, isotopes
    - Bohr's model of the electron cloud, including electron pairs and the octet rule
- Chemical bonds
    - Ionic bond
    - Covalent bond
    - Polar bond
- Intermolecular forces
    - van der Waals forces
    - Polar interactions
    Hydrogen bridges
- Salts
    - Salt formulas
    - IUPAC nomenclature ionic compounds
    - Precipitation reactions
- Chemical calculations/stoichiometry
    - Mole
    - Concentrations
    - Equilibrium concentrations
    - Acid/ base calculations
    - Ideal gas law
- Equilibrium
    - Equilibrium quotient
    - Equilibrium constant
    - Chatelier's principle
- Acid–base theory
    - Bronsted definitions
    - Calculation of pH in weak acid
    - Calculation of pH in buffer
- Redox reactions
    - Definitions reductor/oxidator
    - Use of standard potentials
    - Balancing redox equations

- Organic chemistry
  - Alkanes, alkenes, alkynes
  - Alcohols, aldehydes, ketones
  - Acids
  - IUPAC nomenclature
  - Isomerism
  - Stereoisomerism
  - Simple reactions
- Chemical energy
  - Photosynthesis
  - Glycolysis
  - Fossil fuels
  - Oil industry
- Ethics in chemistry
  - Academic integrity
  - Special values for chemists

These topics form the basic knowledge for students.

More advanced courses are generally chosen by students who take up science, while regular courses are often a prerequisite for medicine, dentistry, architecture and art.

The background of the students as well as reasons students take up a particular course should reflect in the choice of subjects. The points not mentioned earlier are possible applications of their chemistry knowledge that can be discussed with the students, or these can be contexts that students need to learn.

Assignment 9.3

For students focusing on science, it is important they get some idea about their future study curriculum. Chemistry taught at a secondary school level differs quite a bit from the chemistry taught at the university level. If you want to students to make career in the field of chemistry, you must introduce them to interesting topics for future research in chemistry.

For all students it is important they get an idea about the role chemistry plays in our everyday life. Chemistry plays an important role in our everyday life this is actually a better term – not only in medicine but often also in everyday products.

Circular design is becoming more and more important, in order to preserve the world supplies of specific substances. In part this is formulated in the UN Sustainable Development goals.

9.4 Some specific issues in teaching chemistry

9.4.1 Atomic models

Introducing concepts of atomic models to students depends on two factors. The first question to ask is "do I need the atomic model to relate phenomena at the macroscopic scale to the microscale." The second question is whether your students are ready to understand these abstract concepts. Students must have reached Piaget's "Formal Operational Stage." Otherwise, they will not be able to draw logical conclusions from the model.

Most textbooks start upper secondary portions with Rutherford's atomic model. This is logical if you want to introduce concepts such as atomic weight, molecular weight, ions and salts. However, if you decide to postpone these concepts and first start on other subjects such as the introduction of organic chemistry, you can explain atomic models later.

When introducing organic chemistry, you should start with introducing and naming different compounds, beginning with alkanes, alkenes, alkynes and oxygen compounds. You would need to introduce the structural formulas. Isomerism can also be introduced this way. Fossil fuels, oil industry and energy transition can be discussed at this level.

Assignment 9.4

If you start with introducing atomic models, Rutherford's model should be introduced and then Bohr's model, and if you wish the quantum mechanical model. You do not need to do this at the same time however. The key concept in teaching is "the need to know" principle (Bulte, Westbroek, de Jong, & Pilot, 2006), which can be applied here. When details regarding electron clouds need to be understood, you can introduce Bohr's model. Later the quantum mechanical model may be used if needed.

The introduction of the concept of models requires extra attention (Prins, Bulte, van Driel, & Pilot, 2008). Important steps while using the atomic model, such as the motivation and argumentation, underlying the development and revision of the model, need to be discussed.

In this case, it must be made clear that a model is a representation of reality that explains observed phenomena. The successive models are ideal to explain why a model needs to be refined when more data are obtained, making a refinement of the model necessary.

The concept of charge is an important concept that the students need to understand. Coulombic interaction is an important factor in the atomic model and more specific in chemical bonds (Joki, Lavonen, Juuti, & Aksela, 2015). Charge is a difficult concept for students, as they cannot directly experience it. However, it is important that students understand the need for a model. The assignment shown in Figure 9.1 is well suited for students to understand the motivation and development of Rutherford's model.

Several persons have played an important role in the development of the atomic model. Some of these people are:
 * *Democritus of Abdera*
 * *Becquerel*
 * *J.J. Thomson*
 * *Marsden*
 * *Geiger*
 * *Rutherford*
 * *Chadwick*
Choose one of these people and make a poster in which you give arguments why this person should receive the Nobel Prize for Chemistry.
When all posters have been presented, you will vote on the Nobel Prize Winner

---

Figure 9.1: Assignment for learning about the atomic model.

Students find the calculations involving isotopic composition difficult. Here, it is important to stress the relationship between the macroscopic scale and the microscale. You need to demonstrate the relationship between the micro- and macroscopic world; otherwise, the introduction of the model will have no real meaning for the students. One of the options is to discuss the difference between metals and nonmetals using this model. The importance of having an agreed upon atomic mass is of enormous importance, for example, in trade. Another aspect of the isotopic ratio is the difference in ratios depending on the source. In doping for example, this relationship was used to demonstrate that the source of testosterone found in urine was vegetal and thus not natural.

9.4.2 Chemical bonds

Introducing Bohr's model, as a refinement of Rutherford's model, is essential for the concept of chemical bonding. The structure of the electron cloud and the energy levels associated with shells are difficult to understand for students. Within the structure, the idea of electron pairs and the octet rule aids to better understand the structure in the cloud. However, the idea of planetary orbitals should be avoided, as shown Figure 9.2.

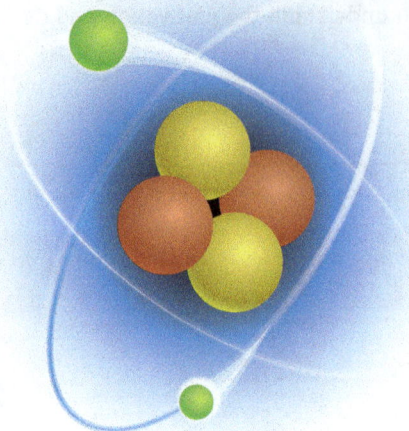

Figure 9.2: Helium with planet orbitals.

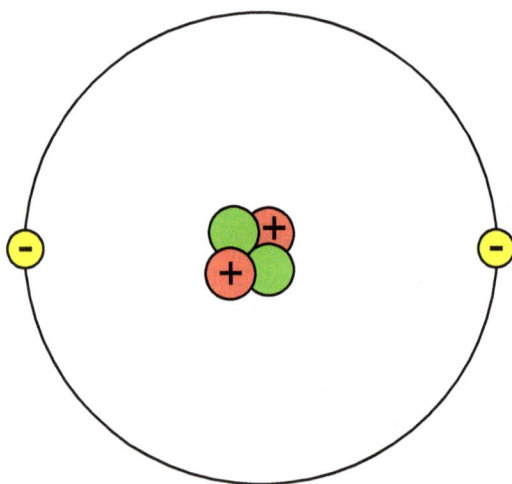

Figure 9.3: Helium with Bohr's shells.

The representation of Figure 9.3 is better.

When you want to refine the model of the electron cloud, going from Bohr's representation to the quantum mechanical one, it is in real terms a refinement or an improvement of the model, while the planetary model is just not correct (see Figure 9.4).

The micro–macro relationship is difficult to understand with chemical bonds. In addition to this, the subtlety of the coulombic interactions is difficult to understand (Joki et al., 2015).

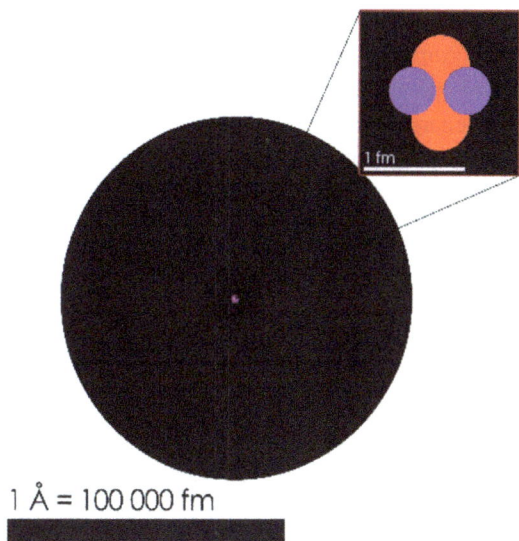

1 Å = 100 000 fm

Figure 9.4: Helium quantum mechanical representation.

Most often the introduction starts with the introduction of the bond between two hydrogen atoms. Next step is the polar bond between hydrogen and chlorine, explaining that the centers of positive and negative charge no longer coincide. There is asymmetrical distribution of charges.

In the bond between sodium and chlorine, this shift of the bonding electron pair to chlorine is complete yielding a positive sodium ion and a negative chlorine ion.

Figure 9.5 shows the charge distribution in the different compounds, making concepts a lot clearer.

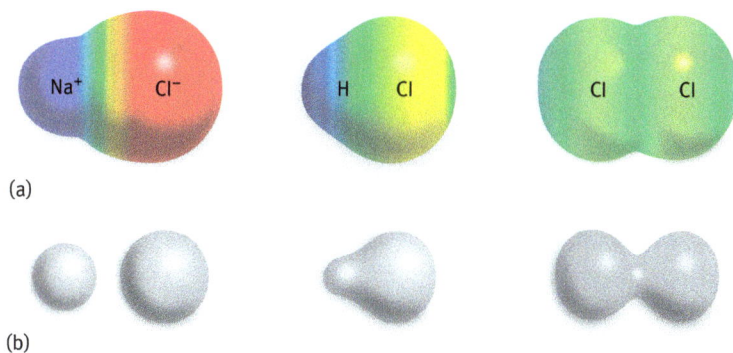

(a)

(b)

Figure 9.5: Charge distribution in Cl_2, HCl and NaCl.

The octet rule can be used to derive the number of bonding electron pairs an element can form. VSEPR (Valence Shell Electron Pair Repulsion) can be used to predict the shape of a molecule. There are several programs that can be used to make such illustrations. In organic chemistry, these charge distributions are important. Without going into quantum mechanics, this is about the limit that can be derived at this level.

9.4.3 Intermolecular forces

The relationship between macroscopic properties and the molecular scale is a lot easier to understand with intermolecular forces. Boiling points and solubility are the most common properties used. It can easily be discussed that these properties have a direct relationship with intermolecular forces. You will need to know about van der Waals forces, polar interactions and hydrogen bridges. These can be directly related to boiling points of various substances. The main problem occurs with the introduction of the concept of hydrogen bridges.

Figure 9.6 is very often used. Figure 9.7, taken from YouTube, gives a much better idea of what happens in liquid water.

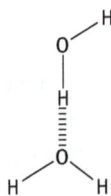

Figure 9.6: Illustration of a hydrogen bridge.

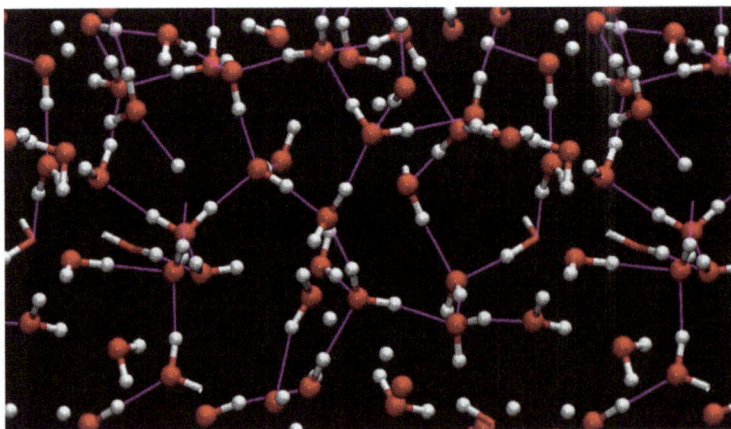

Figure 9.7: Animation of hydrogen bridges: https://youtu.be/Zl74NCVbA5A.

Figure 9.6 shows what is known as dubbed arrow pushing in organic chemistry (Ferguson & Bodner, 2008). In most cases, students are able to draw correct arrows between structural formulas, but have no idea about the meaning of these arrows. They are able to follow the rules but do not understand them. When you introduce hydrogen bridges, it is important to explain the role of these bridges in DNA, RNA and proteins.

9.4.4 Salts

Salts, or ionic compounds, are a traditional subject in the field of chemistry. In the 1950s and 1960s, the H_2S system was part of the curriculum. Solubility rules and solubility calculations are easy to introduce in secondary schools, and students find the concepts of colors of precipitates attractive (Figure 9.8).

Figure 9.8: Chemical garden.

The applications of precipitation reactions at university chemistry level and in everyday life are limited. In treatment plants for drinking and sewage, water precipitation plays a role as well as in pollution by heavy metals.

One of the most difficult concepts for students to understand is the fact that when an ionic compound dissolves in water, the number of particles increases. For some reason, this is contrary to their intuitive knowledge.

9.4.5 Chemical calculations or stoichiometry

The mole is one of the standard units in the SI system of units. The definition of the mole as the number of particles in 12 g $^{12}_{6}C$ is debatable and can change in the near future.

The reason it is contested is the fact that the number of particles depends on the form of carbon. The number of particles in 12 g of diamond is different form the number of particles in 12 g of graphite. In the new definition, Avogadro's number will be considered as the natural constant. The mole will be the amount of substance containing exactly Avogadro's number of particles.

One of the problems that students have with the introduction of the mole is the confusion between the unit "mole" and the amount that belongs to it. According to the gold book (IUPAC, 2017), it was not until 1969 the term "amount of substance" was coined for this amount; before that it was referred to as "the number of moles."

In addition, the letter n has been used in formulas to indicate this particular amount, for example, in the ideal gas law, $pV = nRT$. It is important of indicate which type of particles or entities the amount of substance refers to.

Another controversial issue in stoichiochemistry is the difference between concentration and molarity. The term concentration can indicate different quantities (Cohen et al., 2008), mass concentration, amount (of substance) concentration and number concentration (and volume concentration). Especially when dealing with ionic compounds as well as acids and bases, this difference is of importance. The term molarity, symbol M, is commonly referred to as the amount concentration, symbol c. The number concentration is most often used for particles or entities, symbol C. The square brackets around the symbol for the entity are used as a symbol: $[Cl^-]$.

The unit in all cases is the same $\frac{mol}{dm^3}$, or $\frac{mol}{L}$, often written as $molL^{-1}$. Students find the last notation often confusing.

In science and everyday life, different ways are used to indicate the proportion of a substance. In text books, a number of these are often discussed and are related to the concentrations used in chemistry. These relationships are often difficult to understand, and it may be questioned, for example, what is the use of relating the proof of a liquor to the concentration of ethanol in $molL^{-1}$, even if it is possible to calculate that exactly.

Mass percentages as well as related proportions like ppm and ppb are often used, and in terms of chemical safety they are important units, which should be discussed. It is easier to discuss when talking about chemical safety procedures.

An important issue here is the ways to tackle stoichiometric problems and their calculations. A systematic problem-solving approach will help students solve complicated problems as well. Students need to be trained in using this approach, instead of approaching problems haphazardly or intuitively.

The systematic approach has a few simple steps:
1. Determine which data and information is available.
2. Formulate what needs to be calculated.
3. Determine which links exist between the data and information and what is asked.
4. Formulate a possible way of solving the problem.
5. Carry this possible solution.
6. Evaluate the answer and determine whether the answer fits within the data.

Very often students start with step 3 and skip to 5. For simple problems and exercises, this may work. But in case of complicated problems, students find it difficult.

The last issue is the relationship between the macroscale and the microscale. With these chemical calculations, the switch between the two is often made. It is important to distinguish these steps explicitly for a good understanding by the students.

Assignment 9.5

9.4.6 Equilibrium

Equilibrium can be introduced through physical chemistry by using the following relation:

$$\Delta G^\circ = -RT\log K$$

$G = 0$ is defined as the equilibrium state. Free energy is, however, a difficult concept to introduce in secondary schools. Entropy cannot be easily explained.

The idea that a chemical reaction is reversible and that the direction of a chemical reaction depends on temperature is less difficult. Looking at the formation and decomposition of water, for example, will easily give the idea that there must be some temperature range at which both reactions may occur, and that at some point there will be an equilibrium.

This can be coupled to the notion that at higher temperatures the endothermic reaction occurs, while at lower temperatures the exothermic reaction is favored.

When equilibrium is approached in terms of the rate of reactions, the relationship with concentrations and the reaction quotient "Q" is lot more obvious. It will also be immediately clear what happens when
- $Q < K$
- $Q = K$
- $Q > K$

This way le Chatelier's principle can be introduced in a logical manner.

Students often confuse between Q and K. It is important to stress the difference between the two. Another factor that students find confusing is the difference between "steady state" and equilibrium. At the macroscopic level, this looks more or less the same; however, at the molecular level they are different. The importance and role of equilibrium in concrete situations may help in clarifying the difference.

Equilibrium is normally introduced in the gas phase. The role in solution, however, is important as well. Both precipitation equilibria as well as acid/base equilibria are important aspects in secondary education. In organic chemistry and biochemistry, equilibria are important as well.

9.4.7 Acid/base theory

When introducing acid/base theory, the "need to know" principle (Pilot & Bulte, 2006; Bulte et al., 2006) is an important starting point. Acids and bases are a complicated topic. In a division of IUPAC, professors have been debating the meaning of pH values in oceans. For students, gradually learning about acids and bases works best. Here, linking their own experience of macroscopic phenomena seems a logical pathway. An introduction of organic acids, which give taste to fruits, links to the experience of students. After introducing the organic acids, these acids can easily be linked to inorganic acids by demonstrating the similarity between the two (Figure 9.9).

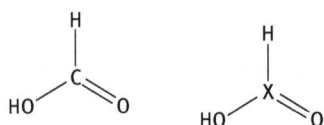

Figure 9.9: Similarity between organic acids and inorganic acids, where X can be any other element.

Linking the macroscopic world to molecules and using symbols to describe these molecules, making sure you are using all angles of Johnstone's triangle.

As illustrated in Figure 9.10 by using organic acids, the hydronium ion can be introduced as the central ion in acids, where $[H_3O^+]$ determines the acidity of a solution. This can be stated more precisely as an equilibrium as shown in Figure 9.11.

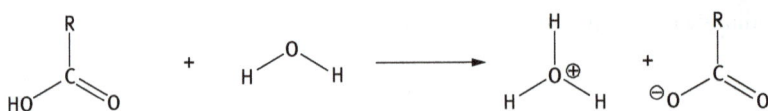

Figure 9.10: Formation of hydronium ion from an organic acid.

Figure 9.11: The formation of hydronium ions written as an equilibrium.

Care must be taken to be precise in the definitions and symbols used. The concentration quotient for the acid equilibrium is given as follows:

$$Q = \frac{[RCOO^-][H_3O^+]}{[RCOOH][H_2O]}$$

This at equilibrium becomes the following:

$$K = \frac{[RCOO^-][H_3O^+]}{[RCOOH][H_2O]}$$

Because $[H_2O]$ is constant, this can be simplified to

$$K_a = \frac{[RCOO^-][H_3O^+]}{[RCOOH]}$$

Discussing this in detail with students will clarify some of the doubts they may have.

One small detail deserves attention, which is the definition of pH.

pH is normally defined as $pH = -\log[H_3O^+]$, as it is not possible to take the log of a unit this should be the numerical value of $[H_3O^+]$.

Once students understand this fact and are able to calculate the pH in a weak acid, expanding to bases is not difficult and Brønsted's theory of acids and bases can easily be introduced, including a more detailed calculation of pH in bases and in buffers.

As students are normally only confronted with phosphate buffers and carbonate/bicarbonate buffers, it is logical to limit the introduction of buffers to these compounds.

9.4.8 Redox reactions

Redox reactions are similar to acid/base reactions where they are both donor–acceptor reactions. This way they can be compared to each other, as is indicated in Table 9.1.

Table 9.1: Similarities between acid/base reactions and redox reactions.

Concept	Acid/base	Redox
Donor	Acid	Reductor
Acceptor	Base	Oxidator
Entity transferred	Proton	Electron
Strength measured relative to	H_2O	H_2/H^+
Strength defined by	K_z/K_b	E_0

The main application of redox reactions is that of batteries. Lithium ion batteries are of special importance for the storage of energy in a time of energy transition. In this sense, the attention has shifted from balancing equation to studying specific applications, such as the hydrogen fuel cell. Basic knowledge about redox reactions is needed to understand these applications. Other applications can be studied as and when needed. After acid/base theory, students usually don't have problems studying redox reactions.

9.4.9 Organic chemistry

In most cases, the study of organic chemistry is divided into two phases. In the first phase, the introduction to nomenclature, alkanes and oxygen compounds is discussed, including some specific topics such as isomerism and stereo isomerism. In the second phase, more specific reactions are discussed, often related to biochemistry, serving as basic knowledge for biochemistry.

Most important problem in the introduction of organic chemistry is that the relationship between macroscopic phenomena and the molecular level is easily lost. It is very easy to stick to formulas and gain knowledge about formulas and to certain extent about the molecules. In this case, students no longer see a connection between, for example, butane and camping gas (Figure 9.12). Relationships to common use of compounds are not always easy to find.

- Organic chemistry
  - Alkanes, alkenes, alkynes
  - Alcohols, aldehydes, ketones
  - Acids
  - IUPAC nomenclature
  - Isomerism
  - Stereoisomerism
  - Simple reactions

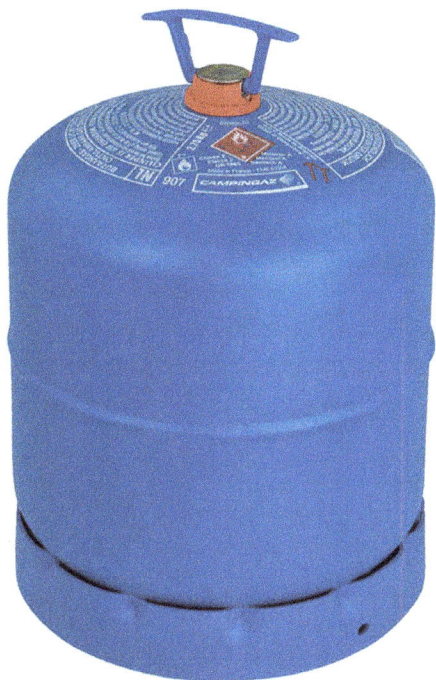

Figure 9.12: Camping gas cylinder filled with butane.

- Chemical energy
  - Photosynthesis
  - Glycolysis
  - Fossil fuels
  - Oil industry

9.4.10 Ethics and chemistry

Academic code of conduct

At some point in time, you will have to discuss an academic code of conduct. As soon as students start to work on more open assignments, they need to be aware of the rules of scientific integrity. Some of the important rules are about plagiarism, cheating, making up or changing data for a project.

Students are normally not aware of the procedures used to properly cite literature. They need to be instructed in the way this works. This should also include some evaluation of the sources as well as the reliability of sources. Facebook and Instagram, for example, cannot be used as a reliable source. Students are generally not aware of the fact that copy–pasting from the internet or other sources is plagiarism, unless properly explained.

Core values for chemists

For chemists, special rules apply. Chemicals can be used for a wide range of purposes. In most cases, these purposes are beneficial. However, whether a particular chemical is beneficial or harmful depends on the context it is used and the intent of its user. For example, 2-propanol is the main ingredient of a cleaning agent. It is also the precursor for mustard gas.

Chemists, basically anybody with enough chemical knowledge, are capable of creating molecules by consciously assembling atoms and molecules into new molecules having specific properties. Everybody working in chemistry, and having knowledge about chemical procedures is not always aware that he/she is not just contributing to fundamental knowledge about chemistry, but also contributes to society, the standard of living and other aspects. This gives chemists a special responsibility towards society.

Core value 1: Chemistry and safety

A chemist is responsible for his safety and that of his colleagues and the environment when working with chemicals and while applying chemical knowledge. Not everybody seems to be convinced of his or her responsibility for themselves or for their environment, as evidenced by the fire on January 5, 2017 in Moerdijk, the Netherlands, of the chemical firm "Chemiepack." On the other hand, industries take their responsibility seriously. After the explosion in a styrene plant, Shell published the cause of the explosion, sharing its knowledge worldwide.

Core value 2: Chemistry and responsibility

A chemist should see to that his/her chemical knowledge is not used for adverse purposes in the society. He should take responsibility not only for his or her work, but also where his/her chemical knowledge is used.

The specialist knowledge a chemist has plays an important role in different levels of society. A chemist knows and recognizes properties of molecules and is well informed about processes related to compounds. The relationship between use and molecular properties of compounds is an important factor. This holds for everybody working with chemicals at any level. It also holds good for dealers of chemical substances as well as the operators working with chemicals in a plant. In addition, it goes for chemistry teachers as well as researchers.

Because chemists have this specific knowledge, they play an important role in the society. With that knowledge, they have responsibilities towards their environment. They can point out potential misuses of chemicals; for example, the production of drugs ("breaking bad"), explosives (terrorism) or chemical weapons. This means that any chemist who suspects misuse or improper use of chemicals or chemical knowledge should take the responsibility to notify the authorities, either directly or through a confidential advisor. Larger companies often have developed procedures for this.

Core value 3: Chemistry and sustainability

A chemist contributes as far as possible to the sustainable use of substances, and so the present use of raw materials has no influence on their use in the future. The amount of raw materials for products is limited. Products can be used as raw materials for new products (cradle to cradle). This means that products after use still have an economical value and can be used as new raw materials. This not only implies adjustment of the production process, but also implies changes in the whole chain of suppliers and users. The influence of production process on the climate is a source of major concern. The EU has formulated the Lund declaration to address the problem. The UN has formulated its sustainable development goals. In Paris in 2016, a new climate treaty was stipulated. In this treaty the amount of energy used in the world played an important role.

Chemists have a special role in reaching these goals because of their specific knowledge. Chemists and chemical knowledge can play a major role in solving the large problems concerning, food, health, drinking water, climate change and energy.

Core value 4: Chemistry and ethics

A chemist working in his/her field should behave ethically and adhere to the prevailing values and standards. This not only concerns his or her own behavior. A chemist is expected to point out unethical behavior by others and report that to the relevant authorities.

Ethics is generally considered to be the doctrine of the difference between good and evil. These concepts are rather different. It deals with norms and standards and about the way and to what extent someone should adhere to these values and standards. Values and standards to a large extent depend on the society in which you are brought up. Within the field of chemistry however, a specific code, about the use of chemical knowledge, combined with the use of chemical substances and technology, is in force. The "Chemical Weapons Convention" in force since 1997 leads to the establishment of OPCW, the organization responsible for removal and destruction of chemical weapons. At the time of writing, only three countries in the world have not yet acceded to this treaty, which sets specific rules for the use and trade of substances as well as technology.

Introducing core values in the classroom

Introducing these core values to students is best done by confronting them a specific situation in which they can apply these concepts. There are many types of applications that can be used as an example:

- Making gamma hydroxy butyric acid, a party drug, which anybody that can make a buffer can make

- The use of "round up" as a weed killer
- The use of plastic microbeads in toothpaste
- Plastic shopping bags

On the internet many examples of consequences of chemical research that are now giving problems can be found. There is a four-step procedure developed at the University of Utrecht (Knippels, Severiens, & Klop, 2009), in which the students are asked specific questions about a case.

These steps are as follows:

Step 1: **Explore**
- Which questions are raised by this specific case?
- What information is missing at this point?
- What is the moral question?
- Which modes of action are evident/possible at first sight?

Step 2: **Analyze**
- Who are involved in this moral dilemma?
- Which arguments are relevant in answering the moral question?

Step 3: **Weigh**
- What is the weight of the arguments of this specific case?
- Based on this weighing, which modes of action can be preferred?

Step 4: **Approach**
- What concrete steps (action) have to be undertaken based on this decision?
- What will be the consequences of this action?

These four steps will structure the discussion for your students and focus on the moral dilemma. For a particular case, if you are not focusing on specific points in the discussion, it will lose the focus from the case intend to present.

Assignment 9.6

Assignment 9.1

Use peer teaching (Mazur, 1997) to design a test for your students to determine whether they have mastered the prerequisite concepts discussed in this section.

For peer teaching you need to design a multiple choice question about a specific concept.

For example,

What can you say about pure mineral water?

Figure A9.1: Perrier mineral water.

It is a pure substance.

It is a mixture.

It is a suspension.

I don't have enough data.

Steps in the process:

Show the question.

Let your students think for themselves for a minute.

By hand gesture take an inventory of the answers.

If 90% answers correct, proceed to the next question.

If answers are divided, give the students time to discuss the answer with their neighbors.

Take another inventory of the answers now.

Discuss the correct answer.

Before you administer the test, discuss it with one of your colleagues or coach this should be coach. After having administered the text discuss the results with the same colleague/coach.

Add the test to your portfolio.

Assignment 9.2

Make an inventory of the curriculum for biology, physics and mathematics related to the science and mathematical skills students are expected to have acquired. Discuss these with the science and mathematics teachers who also teach at your level. Ask them about their experiences about the knowledge and skill levels of the students.

Write a short report, and discuss this with your coach and fellow students.

Add the report to your portfolio.

Assignment 9.3

Choose a chapter in the textbook that you are using.

Identify the learning goals in that chapter for both concepts and skills.

Make a table in which you compare the learning goals in that chapter to the related learning goals in the IChO. Identify which parts of the national curriculum are covered by these learning goals.

Make a table in which you present the differences between the three.

Present the results to your fellow students and coach. Discuss the implications of your study for your teaching.

Add the report to your portfolio.

Assignment 9.4

Before you start discussing atomic models, your students need to understand the concept of charge. They should also have at least an intuitive idea of Coulomb's law: $F = \frac{1}{4\pi\varepsilon_0}\frac{Q_1 Q_2}{r^2}$

Discuss with your physics teacher what a good way would be to refresh the knowledge your students have about charge.

Design a short pre- and posttest to check their knowledge and skills.

Design a 15 min item to introduce charge.

Before doing the introduction let your students take the pretest.

In the lesson after your intervention about charge, administer the posttest.

Compare the two to see how effective your intervention was.

Write a short conclusion and add it to your portfolio.

Assignment 9.5

An example of a problem that illustrates the need for a systematic problem approach is given as follows:

Strawberries contain 15% solids and 85% water. To make strawberry jam, strawberries are mashed and mixed with sugar in a weight ratio of 45:55. The mixture is heated to evaporate some of the water and to pasteurize the jam. The heating is continued until the mixture contains only 33.3% water.

Calculate the amount of strawberries in kilograms needed for 1.00 kg of jam.

Carry out the calculation and show the steps in the systematic problem-solving approach.

Another example is as follows:

The cooling fluid used in a car is generally a mixture of glycol and water. The freezing point of the mixture depends on the glycol concentration. The graph is taken from Glycoshell ("NAP"-free antifreeze with high performance of the "long life" type. Protects the cooling system against rust and frost. Can be used in all types of vehicles. Recommended mixture between 33% and 50% water).

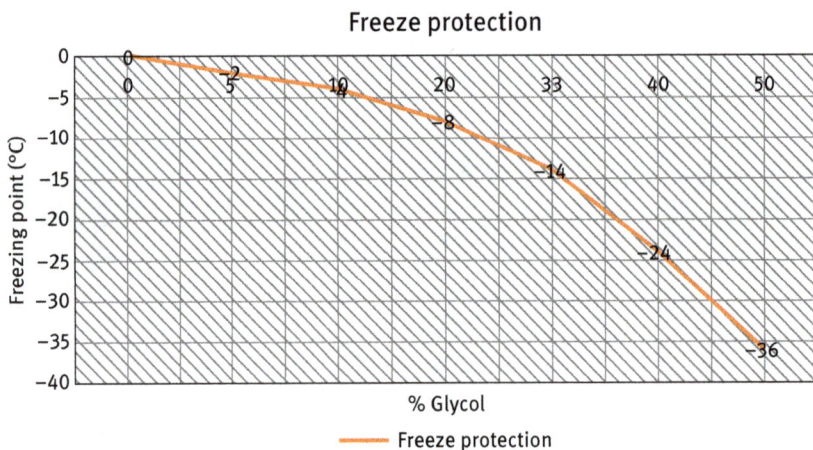

Figure A9.2: Freezing point of glycol.

Suppose that you have 2.5 L cooling liquid, which protects unto −8 °C, and you wish to improve the protection up to −36 °C. How many liters 90% glycol must be added to the liquid?

Add the solutions to your portfolio.

Assignment 9.6

Find a problem you wish to discuss with your students that will fit the four steps mentioned.

Introduce the core values specific for chemists and then present the problem. Let the students discuss the problem using the four steps.

Evaluate the discussion with your students, and see if you can get to some conclusion about action the student could take themselves.

Write a short report and share this with your peers.

Add this to your portfolio.

10 Professional development in chemistry education

10.1 Teaching

As a teacher you may wonder how much influence do you have on the results of your students?

What determines the quality of a teacher?

How can you improve as a teacher?

These are questions related to your professional development as a teacher. Teachers seem to have an influence of between 15% and 25% on the results their students achieve (van de Grift, Chun, Maulana, Lee, & Helms-Lorenz, 2017). Other factors that determine the results of your students lie in the background of the students. As discussed before, students differ in their aptitude, intelligence and motivation. These are all factors important for their success. Social background and gender are other factors. Considering these facts, about 15–25% is a fairly high influence factor.

Instruments have been developed to assess the quality of your teaching, including the impact you have on your students (van de Grift, 2007). These instruments are based on observing the teachers as they teach in the classroom and interact with students during a lesson, normally in a 50-min period. During these observations, a number of aspects need to be considered.

Table 10.1 lists these indicators of the quality of teaching (based on the forms used at the teacher training department in Groningen as well as on van de Grift, 2007).

It takes time to attain the flexibility and insight to see all these indicators in a lesson. Some aspects should be given priority. Without an efficient classroom management, it is not be possible to create a safe and stimulating learning environment. A clear concise instruction is needed to make your teaching effective. To implement learning strategies, easy interaction with your students is also needed.

Self-analysis, which means relating your own actions in the classroom to student's behavior, is one of the most important factor in developing yourself as a teacher. Maintain a log of your teaching activities, in which you note the preparation and lesson plans as well as specify the evaluation of your lesson, can help you improving your teaching skills.

These indicators are valid for any subject taught in the school. They are not specific only for chemistry. However, for chemistry there are a few extra indicators need to be taken into account (Table 10.2), focusing mainly on safety in the classroom and laboratory. As a chemistry teacher you are also responsible for the storage of chemicals as well as chemical glassware and equipment. There are special regulations for using chemical compounds. A number of compounds may no longer be used in school labs. For example, compounds such as tetrachloromethane,

https://doi.org/10.1515/9783110569629-010

Table 10.1: Indicators for the quality of teaching (from Van de Grift (2007).

Indicators	Good practice examples
Efficient classroom management	Clearly recognizable components in the lesson Orderly progression of the lesson Time efficient Clear instructions to students Regularly checks attention level of students
Safe and stimulating learning climate	Relaxed atmosphere in the classroom Encourages mutual respect Demonstrates affinity with social world of the students Supports self-efficacy Stimulates group building Encourages cooperation between students Reacts adequately on disturbances/unwished behavior
Clear instruction	Introduces (learning) goals of the lesson; lesson has a clear start Evaluates the lesson; lesson has a clear ending Can give clear explanations and instructions Checks whether students understood the material Is interactive with the students Is able to activate the students
Differentiative teaching	Is aware of the learning level of his/her students Adjusts his/her lesson in case the learning goals are set too high Is able to improvise, based on the feedback from students Adjusts instructions to the individual student Allows students to work in groups Has extra material for weaker students and stronger students what I mean here is better students, and students that need more time to learn. If you have a better term for weaker and stronger fine
Learning strategies	Relates to prior knowledge Relates to what was learned in other areas Encourages application of learned content Uses systematic problem-solving schemes Teaches students to relate outcomes of problems to practical context Teaches students to simplify problems Teaches students to consult and refer to sources
Activating of students learning	Uses interactive forms of teaching Stimulates discussion among students Uses materials that activate students' learning Permits work in small groups

Table 10.1 (continued)

Indicators	Good practice examples
Self-analysis	Is open to feedback
	Is able to indicate strong and weak aspects of a lesson
	Is able to relate student's behavior, feelings, needs in the analysis
	Is able to relate back to pedagogical theories
	Can formulate improvements
	Is able to formulate positive and negative aspects of alternatives
Observed student's behavior	Students listen attentively to instructions
	Students take active part in classroom activities
	Students are focused and concentrate on their assignments
	Students pose questions and take initiative
Position in the school	Is an active member of the team
	Exchanges information with colleagues
	Is aware of procedures and regulations within the school, and implements these
	Is constructive in his/her contact with parents and guardians of students

Table 10.2: Indicators for chemistry teachers.

Indicators	Good practice examples
Safety in the classroom	Is aware of safety when demonstrating experiments
	Wears lab coat and goggles when demonstrating an experiment
	Is able to explain and maintain safety procedures in the classroom
Safety in the school laboratory	Has a set of safety rules in place
	Sees to it that the lab is properly equipped with safety equipment
	Performs a safety drill at least once a year
	Instructs students at the beginning of a lab on safety
	Students wear lab coats or aprons and goggles
	Designs practical assignments with care
	Gives students responsibility for their own safety

potassium chromate and liquid mercury, which were quite common, are no longer permitted. In addition, the glassware used in secondary-school labs is changing. The lab shown in the Netflix series "Breaking Bad" was exceptionally well equipped for a high-school lab.

It takes 3–5 years to become an experienced teacher. By that time, you would have taught chemistry at all levels in your school. You would have developed your own teaching style and established a specific atmosphere in your classroom, which you and your students find comfortable. Most of the good practices listed in

Table 10.1 hold good for you. This is the reason why in educational research teachers with experience less than 5 years are called "novice," 5–9 years "early career," 10–18 years "mid-career," and over 18 years "late career" (Davidowitz & Potgieter, 2016)

After 3–5 years, you become aware of difficulties students face in specific topics in chemistry, thus helping you improve your pedagogical content knowledge (PCK). (see Figure 10.1)

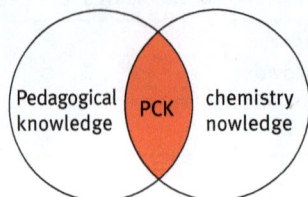

Figure 10.1: Pedagogical content knowledge.

PCK can be considered as the intersect of knowledge with regard to teaching or pedagogy and chemistry content knowledge. During your career, you learn more about pedagogy through experience. With time you gain insight into the specific problems that students have with learning chemical concepts. PCK is widely researched with regard to chemistry education (Loughran, Mulhall, & Berry, 2004).

The idea about PCK has been slightly refined, as is shown in Figure 10.2, linking it more to the development of teachers' knowledge (Gess-Newsome, 2015).
As a teacher you start building up your knowledge base about assessment, content and students' curriculum. You can use this for a specific topic, together with your

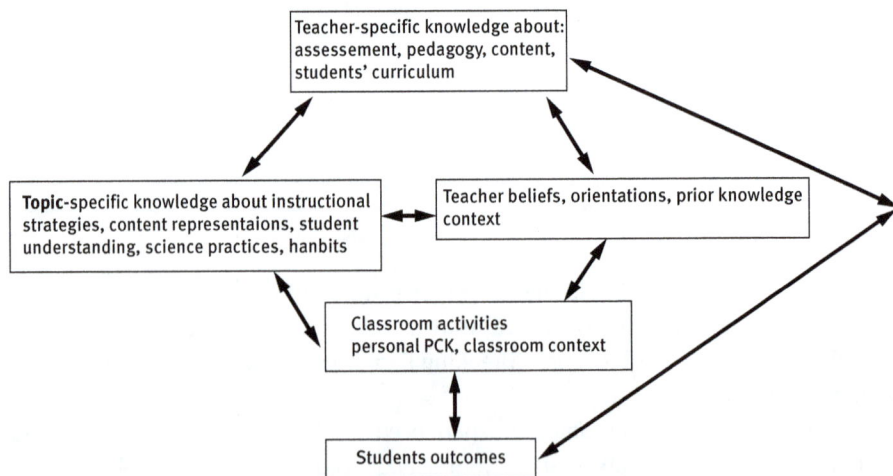

Figure 10.2: Professional teachers' knowledge. From Gess-Newsome (2015).

own ideas and experiences to develop classroom activities, which lead to outcomes. These outcomes along with the classroom experiences become part of your knowledge base. You can use outside knowledge and experience in your own teaching. Other people's experience can also help you and can give you some insight in trying out new activities in the classroom, or find another context to use in your teaching.

EBSCO is an easy search module to find the literature. By "google scholar" a number of other sources may be found. Unfortunately, all material is still not open for access. However, through the university where you were trained, it is often possible to gain access to this type of search engines and literature.

Generally, your own science or chemistry teacher association will have a journal in which articles are published regularly. For example, ACS has its *Journal of Chemical education*; RSC has several journals, including *Education in Chemistry* and the Association of Science Education has *School Science Review*. A new international journal is *Chemistry Teacher International*, published by IUPAC and DeGruyter.

Improving your teacher-specific knowledge is a process that requires you learn from and cooperate with others, which is your team of chemistry teachers in your school. Very often there are only a few chemistry teachers in secondary schools. Generally, there will be an university in your area that does research in the field of chemistry education. The groups doing research in chemistry education normally have a network of teachers they work with. You can become part of such a network and work together on further development of your educational skills and experience. Again here the institution where you were trained as a teacher can help you find a network you can link to.

An important aspect for your professional development is maintaining a log of your activities. If you want to discuss something that happened in the classroom with your colleagues, the notes would help. Figure 10.3 shows an example of such a log.

It does not take time to make these notes. Together with the material you used in the classroom, it will help you analyze the result of your classroom activities. It will also help you share your ideas and experiences with others in your network.

More importantly, it will help you evaluate the way you have taught a chapter. It is probable you can link results on a final test, administered to your students and the lesson notes you took for a complete evaluation.

An important aspect that you should keep in mind when trying out new activities is that you need to give yourself room to fail. Activities that do not work the first time may work a lot better in a second or third time, after required changes have been made. You need to give yourself some room to learn, and students also often need some time to adjust.

One way to improve yourself as a teacher is to visit the locally, nationally and internationally organized conferences about chemistry education. By visiting the lectures, workshops and activities, you can interact with colleagues, but more importantly you get new ideas and content for teaching.

Lesson date	Lesson time	class	room

Lesson plan

Topic	
Book section	
Learning goals	
Exercises	
Activity 1	
Activity 2	
Activity 3	
Home work	

Lesson notes:

Topic	
Book section	
Learning goals	
Exercises	
Activity 1	
Activity 2	
Activity 3	
Home work	

Figure 10.3: Example of a log book page.

Assignment 10.1

10.2 Educational Research

You can of course get involved in educational research. You can cooperate with researchers, but you can do the research yourself as well. Doing research takes time, but will be rewarding. You may go for a PhD, but at the very least you can publish your results. One of the main differences from regular teaching is that you focus on a particular aspect of teaching, trying to determine the effectiveness, for example, of a specific intervention. Doing research requires a more rigid framework of administration than regular teaching. Normally you work together with two or three people. The easiest way to get started is to seek cooperation from an institution near you that does educational research. In most cases they are more than willing to help you.

Assignment 10.2

10.3 Other tasks within the school

Within school several tasks are carried out by teachers. These teachers allot time to perform these specific tasks, for example, Coordinating a year group, organizing the teachers meeting, representing the teachers on the parent teacher association, organizing excursions and student counselling. Often teachers are asked to become part of the school management. Normally you need at least 5 years of experience as a teacher before you are asked for one of these tasks. It helps you diversify.

Assignment 10.3

Assignment 10.1 Chemistry teacher conferences

During your training, you should try to attend at least one national chemistry teacher conference. There is a special fee for apprentice teachers. You can write a short report about the workshops and lectures you attended, including what you learned and how you can use your learning in your future teaching or what you will not use in your future teaching.

Discuss with your fellow students and add to your portfolio.

Assignment 10.2 Research

Design a short research item, with the help of your fellow students, where you formulate a clear research question that may be answered by carrying out one or two interventions.

Gather the material in a short article.

See if your and your fellow students' efforts can be combined.

Add the report to your portfolio.

Assignment 10.3 Excursion

Participate in an excursion organized within the science department.

Reflect on the added value of such an excursion and decide whether it is worth the effort.

Add the reflection to your portfolio.

Answer to assignment 6.1: about 0.5 g. This implicates that the sublimation of iodine should not be carried out by students in the lab.

References

Abrahams, I., & Millar, R. (2008). Does practical work really work? A study of the effectiveness of practical work as a teaching and learning method in school science. *International Journal of Science Education*, *30*(14), 1945–1969. https://doi.org/10.1080/09500690701749305

Anderson, L. W., & Krathwohl, D. R. (2001). *A Taxonomy for learning, teaching, and assessing : a revision of Bloom's taxonomy of educational objectives* (Abridged e). New York; Longman.

Anderson, L. W., & Krathwol, D. R. (2001). The knowledge dimension. In L. W. Anderson & D. R. Krathuol (Eds.), *A taxonomy for learning, teaching and assesing (a revision of Bloom's taxonomy of educational objectives*. New York: Addison Wesley Longman.

anonymous. (2012). workprogramme 2013 capacities part 5, science in society. Brussels: EU. Retrieved from http://ec.europa.eu/research/participants/data/ref/fp7/134006/s-wp-201301_en.pdf

Apotheker, J. (2018). The irresistible use of contexts in chemistry education. *Israel Journal of Chemistry*.

Apotheker, J. H. (2004). Viervlakkig Chemie Onderwijs. *NVOX*, *29*(9), 488–490.

Apotheker, J. H. (2005). De isolatie van carvon. *NVOX*, *30*(1), 13–14.

Apotheker, J. H. (2006). Chik in Groningen. *Ilka, submitted*.

Apotheker, J. H. (2008). Introducing a context based curriculum in the Netherlands. In *20th International Conference on Chemical Education*. Mauritius: Caslon printing ltd, Mauritius.

Apotheker, J. H. (2009). Context and Chemistry Going Dutch? The Development of a Context-Based Curriculum in the Netherlands. In *Chemistry Education in the ICT Age* (pp. 119–129). Dordrecht: Springer. https://doi.org/10.1007/978-1-4020-9732-4_14

Apotheker, J. H., Bulte, A., de Kleijn, E., van Koten, G., Meinema, H., & Sellar, F. (2010). *Scheikunde in de dynamiek van de toekomst, Eindrapport van de stuurgroep Nieuwe Scheikunde 2004–2010*. Enschede: SLO.

Apotheker, J. H., & Teuling, E. (2017). *Carbohydrates in breastmilk*. Kiel: IPN.

Astrophysics, H.-S. C. for. (1997). Minds of our own. USA.

Bertona, C., Kleijn, E. de, Hennink, D., Apotheker, J. H., Drooge, H. van, Waals, M., … Lune, J. van. (2014). *Scheikunde VWO, Syllabus Centraal Examen 2016*. Utrecht: College voor examens.

Biggs, J. (1996). Enhancing teaching through constructive alignment. *Higher Education*, *32*(3), 347–364.

Bijsterbosch, E., van der Schee, J., & Kuiper, W. (2017). Meaningful learning and summative assessment in geography education: an analysis in secondary education in the Netherlands. *International Research in Geographical & Environmental Education*, *26*(1), 17–35. Retrieved from http://10.0.4.56/10382046.2016.1217076

Black, P., & Wiliam, D. (1998). Inside the black box. *Phi Delta Kappa*, *80*(2), 139–44.

Blackman, A., Bottle, S., Schmid, S., Mocerino, M., & Wille, U. (2011). *Chemistry* (2nd ed.). Frankfurt: Wiley.

Bloom, B. S. (1984). *Taxonomy of educational objectives: Book 1 cognitive domain*. Reading MA: Addison Wesley.

Borley, M., Harden, H., Gardom Hulme, P., Palmer, E., Tiernan, A., Warren, D., & Dunlop, L. (2016). *Twenty first century science: chemistry for GCSE combined science student book* (3rd ed.). Oxford: OUP. Retrieved from https://global.oup.com/education/content/secondary/series/21st-century-science-3ed/?view=ProductList&region=international

Bransford, J. D, Brown, A. L., & Cocking, R. (2000). *How people learn*. Washington D.C.: National Academy Press.

Broadfoot, P., Daughtery, R., Gradner, J., Harlen, W., James, M., & Stobart, G. (2012). assessment for learning. Retrieved November 21, 2017, from https://assessmentreformgroup.files.wordpress.com/2012/01/10principles_english.pdf

https://doi.org/10.1515/9783110569629-011

Broadfoot, P., Gardner, J., Daugherty, R., & Gipps, C. (1999). *Assessment for learning: beyond the blackbox.* Cambridge: University of Cambridge, School of Education.

Buck, L. L., Bretz, S. L., & Towns, M. H. (2008). Characterizing the level of inquiry in the undergraduate laboratory. *Journal of College Science Teaching, 38*(1), 52–58.

Bulte, A. M. W., Westbroek, H. B., de Jong, O., & Pilot, A. (2006). A research approach to designing chemistry education using authentic practices as contexts. *International Journal of Science Education, 28*(9), 1063–1086. Retrieved from http://search.ebscohost.com.proxy-ub.rug.nl/login.aspx?direct=true&db=eric&AN=EJ740453&site=ehost-live&scope=site

Butcher, D. J., Brandt, P. F., Norgaard, N. J., Atterholt, C. A., & Salido, A. L. (2003). Sparky IntroChem: a student-oriented introductory chemistry course. *Journal of Chemical Education, 80*(2), 137–139.

Bybee, R. W., Powell, J. C., & Towbridge, L. (2007). *Teaching secondary school science, strategies for developing science literacy* (9th ed.). Upper Saddle river, N.J.: Pearrson.

Bybee, R. W., Taylor, J. A., Gardner, A., Van Scotter, P., Powell, J. C., Westbrook, A., & Lamdes, N. (2006). *The BSCS 5E instructional model: origens and effectiveness.* Colorado Springs, CO: BSCS.

Carey, J., Churches, R., Hutchinson, G., Jones, J., & Tosey, P. (2010). *Neuro-linguistic programming and learning: teacher case studies on the impact of NLP in education.* Online Submission. ERIC. Retrieved from http://search.ebscohost.com.proxy-ub.rug.nl/login.aspx?direct=true&db=eric&AN=ED508368&site=ehost-live&scope=site

Chan, C. (2006). Teaching the Chinese learner in higher education. Retrieved April 19, 2018, from http://media.leidenuniv.nl/legacy/2Historic_and_cultural_background.pdf

Cohen, E. G. (1994). Restructuring the classroom: conditions for productive small groups. *Review of Educational Research, 64*, 1–35.

Cohen, E. R., Cvitas, T., Frey, J. G., Holmström, B., Kuchitsu, K., Marquardt, R., … Thor, A. J. (2008). *Quantities, units and symbols in physical chemistry, IUPAC Green Book* (3rd ed.). Cambridge: IUPAC and RSC Publishing.

Commission, E. (2017). Work programme 2018–2020. Retrieved March 23, 2018, from http://ec.europa.eu/research/participants/data/ref/h2020/wp/2018-2020/main/h2020-wp1820-swfs_en.pdf

Davidowitz, B., & Potgieter, M. (2016). Use of the Rasch measurement model to explore the relationship between content knowledge and topic-specific pedagogical content knowledge for organic chemistry. *International Journal of Science Education, 38*(9), 1483–1503. Retrieved from http://search.ebscohost.com.proxy-ub.rug.nl/login.aspx?direct=true&db=eric&AN=EJ1106801&site=ehost-live&scope=site

Day, M. C. (1981). Thinking at Piaget's stage of formal operations. *Educational Leadership, 39*(1), 44. Retrieved from http://search.ebscohost.com.proxy-ub.rug.nl/login.aspx?direct=true&db=aph&AN=7734340&site=ehost-live&scope=site

Demuth, R., Parchmann, I., & Ralle, B. (2006). *Chemie in Kontext, Kontexte, Medien, Basiskonzepte, Sekundarstufe II.* Berlin: Cornelsen Verlag.

Domin, D. S. (2009). Considering laboratory instruction through Kuhn's view of the nature of science. *Journal of Chemical Education, 86*(3), 274. https://doi.org/10.1021/ed086p274

Earle, S., & Davies, D. (2014). Assessment without levels. *Education in Science,* (258), 30–31. Retrieved from http://search.ebscohost.com.proxy-ub.rug.nl/login.aspx?direct=true&db=eric&AN=EJ1064574&site=ehost-live&scope=site

Eisenkraft, A. (2003). Expanding the 5E model. *The Science Teacher, 60* (September), 57–58,59.

Eisenkraft, A., & Freebury, G. (2003). *Active chemistry.* Armonk, NY: Its about time.

exploratorium. (2006). Raising questions, a professional development curriculum from the Institute for Inquiry. San Fransisco: exploratorium. Retrieved from http://www.exploratorium.edu

Fahlman, B. D., Purvis-Roberts, K. L., Kirk, J. S., Bentley, A. K., Daubenmire, P. L., Ellis, J. P., & Mury, M. T. (2017). *Chemistry in context* (9th ed.). New York, N.Y.: McGraw-Hill Education.

Ferguson, R., & Bodner, G. M. (2008). Making sense of the arrow-pushing formalism among chemistry majors enrolled in organic chemistry. *Chemistry Education Research and Practice*, *9*(2), 102–113. Retrieved from http://search.ebscohost.com.proxy-ub.rug.nl/login.aspx?direct=true&db=eric&AN=EJ888327&site=ehost-live&scope=site

French Embassy NY French,Cultural Services, N. Y. (1971). *Education in France: number 41.* Retrieved from http://search.ebscohost.com.proxy-ub.rug.nl/login.aspx?direct=true&db=eric&AN=ED046301&site=ehost-live&scope=site

Galilei, G. (2001). *Dialogue concerning the two chief world systems.* (S. Gould, Ed.). New York, N.Y.: the modern library.

Gardner, H. (1993). *Frames of mind, the theory of multiple intelligences.* London: Fontana press.

Geboers, E., Geijsel, F., Admiraal, W., Jorgensen, T., & ten Dam, G. (2015). Citizenship development of adolescents during the lower grades of secondary education. *Journal of Adolescence*, *45*, 89–97. https://doi.org/10.1016/j.adolescence.2015.08.017

Gejda, L. M., & LaRocco, D. J. (2006). *Inquiry-based instruction in secondary science classrooms: a survey of teacher practice paper presented at the 37th Annual Northeast Educational Research Association Conference (Kerhonkson, NY, Oct 18– 20,2006). Online Submission.* Kerhonkson, New York: ERIC. Retrieved from http://search.ebscohost.com.proxy-ub.rug.nl/login.aspx?direct=true&db=eric&AN=ED501253&site=ehost-live&scope=site

Gess-Newsome, J. (2015). A model of teacher professional knowledge and skill including PCK: results of the thinking from the PCK Summit. In A. Berry, P. Friedrichsen, & J. Loughran (Eds.), *Re-examining pedagogical content knowledge in science education* (pp. 28–42). New York, N.Y.

Gilbert, J. K. (2006). On the nature of "context" in chemical education. *International Journal of Science Education*, *28*(9), 957–976. Retrieved from http://search.ebscohost.com.proxy-ub.rug.nl/login.aspx?direct=true&db=eric&AN=EJ740457&site=ehost-live&scope=site

Gray, P. (2008). a brief history of education. Retrieved April 19, 2018, from https://www.psychologytoday.com/us/blog/freedom-learn/200808/brief-history-education

Groot, A. D. de. (1946). *Het denken van den schaker.* University of Amsterdam.

Heacox, D. (2002). *Differentiating instruction in the regular classroom.* Minneapolis: Free spirit publishing inc.

Heikkinen, H. (2002). *Chemistry in the community* (4th ed.). New York: W.H. Freeman and company.

Hein, G. E. (1975). The social history of open education: Austrian and Soviet schools in the 1920s. *The Urban Review*, *8*(2), 96–119. https://doi.org/10.1007/BF02208898

Hemraj-Benny, T., & Beckford, I. (2014). Cooperative and inquiry-based learning utilizing art-related topics: teaching chemistry to community college nonscience majors. *Journal of Chemical Education*, *91*(10), 1618–1622. Retrieved from http://10.0.3.253/ed400533r

Hofstein, A., & Lunetta, V. N. (1982). The role of the laboratory ins cience teaching: negelcted aspects of research. *Review of Educational Research*, *52*(2), 201–217.

Hooijmaaijers, T. (2000). *Bron en Perspectief, Het belang van Jenaplan, Montessori, Dalton, Freinet en de Vrije School voor het basisonderwijs. Pabo-katernen.* Heeswijk-Dinther: Esstede.

IUPAC. (2017). Goldbook.

Johnson, R.T., Johnson, D. W. (1999). *Learning together and alone* (5th ed.). Boston: Allyn and Bacon.

Johnson, D. W., & Johnson, R. T. (1989). *Cooperation and competition: theory and research* (2nd ed.). Edina, Minn: Interaction book.

Johnstone, A. H. (1997). Chemistry teaching: science or alchemy. *Journal of Chemical Education*, *74*(3), 262–268.

Joki, J., Lavonen, J., Juuti, K., & Aksela, M. (2015). Coulombic interaction in Finnish Middle School Chemistry: a systemic perspective on students' conceptual structure of chemical bonding. *Chemistry Education Research and Practice*, *16*(4), 901–917. Retrieved from http://search. ebscohost.com.proxy-ub.rug.nl/login.aspx?direct=true&db=eric&AN=EJ1077316&site=ehost-live&scope=site

Jong, T. de. (1996). Types and qualities of knowledge. *Educational Psychologist*, *31*(2), 10–113.

Kagan, S. (1990). The structural approach to cooperative learning. *Educational Leadership*, *47*(4), 12–15.

Kansanen, P. (2003). Studying the realistic bridge between instruction and learning. An attempt to a conceptual whole of the teaching-studying-learning process. *Educational Studies (03055698)*, *29* (2),221. Retrieved from http://search.ebscohost.com.proxy-ub.rug. nl/login.aspx?direct=true&db=aph&AN=10282649&site=ehost-live&scope=site

Kansanen, P., & Merk, M. (1999). Kansanen, P. & Meri, M. (1999). The didactic relation in the teaching-studying-learning process, in B. Hudson, F. Buchberger, P. Kansanen & H. Seel (Eds) Didaktik/Fachdidaktik as Science(-s) of the Teaching Profession, pp. 107–116 (TNTEE Publications). In B. Hudson, F. Bushberger, P. Kansanen, & H. Seel (Eds.), *Didaktik/ Fachdidaktik as Science(-s) of the teaching professions* (pp. 107–116). Helsinki: TNTEE.

Kimberlin, S., & Yezierski, E. (2016). Effectiveness of inquiry-based lessons using particulate level models to develop high school students' understanding of conceptual stoichiometry. *Journal of Chemical Education*, *93*(6), 1002–1009. Retrieved from http://10.0.3.253/acs.jchemed.5b01010

King, D. (2012). New perspectives on context-based chemistry education: using a dialectical sociocultural approach to view teaching and learning. *Studies in Science Education*, *48*(1), 51–87. Retrieved from http://search.ebscohost.com.proxy-ub.rug.nl/login.aspx? direct=true&db=eric&AN=EJ957174&site=ehost-live&scope=site

Kirschner, P. A., Sweller, J., & Clark, R. E. (2006). Why minimal guidance during instruction does not work: an analysis of the failure of constructivist, discovery, problem-based, experiential, and inquiry-based teaching. *Educational Psychologist*, *41*(2), 75–86. https://doi.org/10.1207/ s15326985ep4102_1

Knippels, M. P. J., Severiens, S. E., & Klop, T. (2009). Education through fiction: acquiring opinion-forming skills in the context of genomics. *International Journal of Science Education*, *31*(15), 2057–2083. https://doi.org/10.1080/09500690802345888

Konak, A., Clark, T. K., & Nasereddin, M. (2014). Using Kolb's Experiential Learning Cycle to improve student learning in virtual computer laboratories. *Computers & Education*, *72*, 11–22. https://doi.org/10.1016/j.compedu.2013.10.013

Kruit, P. M., Oostdam, R. J., van den Berg, E., & Schuitema, J. A. (2018). Effects of explicit instruction on the acquisition of students' science inquiry skills in grades 5 and 6 of primary education. *International Journal of Science Education*, *40*(4), 421–441. Retrieved from http://search.ebscohost.com.proxy-ub.rug.nl/login.aspx?direct=true&db=eric&AN=EJ1172 738&site=ehost-live&scope=site

Kuri, N. P. (2000). *Kolb's Learning Cycle: an alternative strategy for engineering education.* Retrieved from http://search.ebscohost.com.proxy-ub.rug.nl/login.aspx?direct=true&db=eric &AN=ED441666&site=ehost-live&scope=site

Laherto, A., Kampschulte, L., de Vocht, M., Blonder, R., Akaygun, S., & Apotheker, J. (2018). Contextualizing the EU's "Responsible Research and Innovation" policy in science education: a conceptual comparison with the nature of science concept and practical examples. *EURASIA Journal of Mathematics, Science & Technology Education*, *14*, 1–15. https://doi.org/10.29333/ ejmste

Le Hebel, F., Montpied, P., Tiberghien, A., & Fontanieu, V. (2017). Sources of difficulty in assessment: example of PISA science items. *International Journal of Science Education*, *39*(4),

468–487. Retrieved from http://search.ebscohost.com.proxy-ub.rug.nl/login.aspx?
direct=true&db=eric&AN=EJ1136369&site=ehost-live&scope=site

Likert, R. (1978). *A technique for the measurement of attitudes*. Univesity of Michican, Ann Arbor Michigan.

Lionni, L. (1970). *Fish is fish*. New York: dragonfly books.

Loughran, J., Berry, A., & Mulhall, P. (2006). *Understanding and developing science teachers' pedagocical content knowledge*. Rotterdam: Sense Publishers.

Loughran, J., Mulhall, P., & Berry, A. (2004). In search of pedagogical content knowledge in science: developing ways of articulating and documenting professional practice. *Journal of Research in Science Teaching*, *41*(4), 370–391.

Mahaffy, P. (2004). Chemistry education: the shape of things to come. In *18th ICCE* (p. 2). Istanbul.

Marzano, R. J., & Association for Supervision and Curriculum Development VA., A. (1992). *A different kind of classroom: teaching with dimensions of learning*. Retrieved from http://search.ebscohost.com.proxy-ub.rug.nl/login.aspx?direct=true&db=eric&AN=ED 350086&site=ehost-live&scope=site

Mayer, R. E. (2009). *Multimedia learning* (2nd ed.). Cambridge: Cambridge University Press.

Mazur, E. (1997). *Peer instruction: a user's manual*. Upper Saddle river, NJ: Prentice Hall.

Mcleod, S. A. (2014). Lev Vygotsky.

McPherson, H. (2018). Transition from cookbook to problem-based learning in a high school chemistry gas law investigation. *Teaching Science: The Journal of the Australian Science Teachers Association*, *64*(1), 47–51. Retrieved from http://search.ebscohost.com.proxy-ub.rug. nl/login.aspx?direct=true&db=aph&AN=128835416&site=ehost-live&scope=site

Miller, D., & Kandl, T. (1991). Knowing … what, how, why. *Australian Mathematics Teacher*, *47*(3), 4–8. Retrieved from http://search.ebscohost.com.proxy-ub.rug.nl/login.aspx?
direct=true&db=eric&AN=EJ446389&site=ehost-live&scope=site

Montessori, M. (1912). *The montessori method* (20th ed.). New York: Schocken books.

Mullis, I. V. ., & Martin, M. O. (2015). *TIMSS 2015 assessment frameworks*. Boston. Retrieved from https://timssandpirls.bc.edu/timss2015/frameworks.html

Nations, U. (2016). Transforming our world: the 2030 Agenda for Sustainable Develeopment Knowledge Platform. Retrieved March 18, 2018, from https://sustainabledevelopment-un-org.
proxy-ub.rug.nl/post2015/transformingourworld.

Nentwig, P. M., Demuth, R., Parchmann, I., Gräsel, C., & Ralle, B. (2007). Chemie im Kontext: situated learning in relevant contexts while sytematically developing basic chemical concepts. *Journal of Chemical Education*, *84*(9), 1439–1444.

Nentwig, P. M., Parchmann, I., Gräsel, C., & Ralle, B. (2007). Chemie im Kontext: situated learning in relevant contexts while systematiccaly developing basic chemical concepts. *Journal of Chemical Education*, *84*(9), 1440–1444.

Nick, S., & Nather, C. (2007). Analysis of a superconductor: development of a practical exam for the international chemistry olympiad. *Journal of Chemical Education*, *84*(12), 1951–1954. Retrieved from http://search.ebscohost.com.proxy-ub.rug.nl/login.aspx?direct=true&db=eric&AN=EJ82 0894&site=ehost-live&scope=site

Nurrenbern, S. C., & Pickering, M. (1987). Concept learning versus problem solving: is there a difference. *Journal of Chemical Education*, *64*(6), 508–510.

Oers, B. van. (1998). From context to contextualizing. *Learning and Instruction*, *8*(6), 473–488.

Osborne, J., & Dillon, J. (2008). *Science education in Europe: critical reflections a report to the Nuffield Foundation*. London: The Nuffield Foundation.

Özalp, D., & Kahveci, A. (2015). Diagnostic assessment of student misconceptions about the particulate nature of matter from ontological perspective. *Chemistry Education Research and*

Practice, *16*(3), 619–639. Retrieved from http://search.ebscohost.com.proxy-ub.rug.nl/login.aspx?direct=true&db=eric&AN=EJ1067731&site=ehost-live&scope=site

Pilot, A., & Bulte, A. M. W. (2006). Why do you "need to know"? context-based education." *International Journal of Science Education*, *28*(9), 953–956.

Pota, V. (2017). The future of education: innovations needed to meet the sustainable development goals. *Childhood Education*, *93*(5), 368–371. Retrieved from http://search.ebscohost.com.proxy-ub.rug.nl/login.aspx?direct=true&db=eric&AN=EJ1154468&site=ehost-live&scope=site

Powers, A., Langdon, L., Pentecoast, T., & Schwennsen, C. (2011). *chemistry in the community* (sixth). Washington D.C.: W.H. Freeman and Company.

Prescott, J. O. (1999). A day in the life of the Rudolf Steiner School. *Instructor*, *109*(4), 21–25. Retrieved from http://search.ebscohost.com.proxy-ub.rug.nl/login.aspx?direct=true&db=eric&AN=EJ598272&site=ehost-live&scope=site

Prins, G. T., Bulte, A. M. W., van Driel, J. H., & Pilot, A. (2008). Selection of authentic modelling practices as contexts for chemistry education. *International Journal of Science Education*, *30*(14), 1867–1890. Retrieved from http://search.ebscohost.com.proxy-ub.rug.nl/login.aspx?direct=true&db=eric&AN=EJ815823&site=ehost-live&scope=site

Regulations of the International Chemistry Olympiad. (2013). Retrieved October 10, 2018, from https://50icho.eu/wp-content/uploads/2017/12/Regulations.pdf

Reillon, V. (2017). EU Framework Programmes for Research and Innovation: evolution and key data from FP1 to Horizon 2020 in view of FP9. Retrieved March 23, 2018, from http://www.europarl.europa.eu/thinktank/en/document.html?reference=EPRS_IDA(2017)608697

Rocard, M., Csermely, P., Jorde, D., Lenzen, D., Walberg-Henriksson, H., & Hemmo, V. (2007). *Science education now: a renewed pedagogy for the future of Europe* (Vol. EUR 22845). Brussels: European Commission.

Rocha de dos Reis, P., Marques, A. R., & Azinhaga, P. (2015). *Irresistible exhibitions guide book*. Lisboa.

Rubistar. (n.d.). Retrieved November 16, 2018, from http://rubistar.4teachers.org/index.php?screen=NewRubric

Schultz, T. W. (1989). Investing in people: schooling in low income countries. *Economics of Education Review*, *8*(3), 219–223. Retrieved from http://search.ebscohost.com.proxy-ub.rug.nl/login.aspx?direct=true&db=eric&AN=EJ397756&site=ehost-live&scope=site

Serrel, B., & Raphling, B. (1992). computers on the exhibit floor. *Curator*, *35*(3), 181–189.

Shaw, R. (2013). The implications for science education of Heidegger's philosophy of science. *Educational Philosophy and Theory*, *45*(5), 546–570. Retrieved from http://search.ebscohost.com.proxy-ub.rug.nl/login.aspx?direct=true&db=eric&AN=EJ1011320&site=ehost-live&scope=site

Silberberg, M., & Amateis, P. (2015). *Chemistry, The Molecular Nature of Matter and Change* (7th ed.). New York: McGraw-Hill.

Sirota, A. (2015). The competion problems from the International Chemstry Olympiads volume 1,2 3. Retrieved March 23, 2018, from https://www.iuventa.sk/en/Subpages/ICHO/Past-Competition-Problems.alej

Sjøberg, S., & Schreiner, C. (2010). *The rose project. An overview and key findings*. Oslo. Retrieved from http://www.roseproject.no/publications/english-pub.html

Slooter, M. (2009). *de vijf rollen van de leraar*. Amersfoort: CPS.

Smith, G. (2007). How does student performance on formative assessments relate to learning assessed by exams? *Journal of College Science Teaching*, *36*(7), 28–34. Retrieved from http://search.ebscohost.com.proxy-ub.rug.nl/login.aspx?direct=true&db=eric&AN=EJ769007&site=ehost-live&scope=site

SWELLER, J., KIRSCHNER, P. A., & CLARK, R. E. (2007). Why minimally guided teaching techniques do not work: a reply to commentaries. *Educational Psychologist*, *42*(2), 115–121. https://doi.org/10.1080/00461520701263426

Talanquer. (2014). Chemistry in past and new science frameworks and standards: gains, losses, and missed opportunities. *Journal of Chemical Education*, *91*(1), 24–29.

Toplis, R. (2012). Students' views about secondary school science lessons: the role of practical work. *Research in Science Education*, *42*(3), 531–549. Retrieved from http://search.ebscohost.com.proxy-ub.rug.nl/login.aspx?direct=true&db=eric&AN=EJ964826&site=ehost-live&scope=site

Tytler, R., & Australian Council for Educational Research, V. (2007, January 1). Re-imagining science education: engaging students in science for Australia's future. *Australian Education Review* 51. Australian Council for Educational Research. Retrieved from http://search.ebscohost.com.proxy-ub.rug.nl/login.aspx?direct=true&db=eric&AN=ED499154&site=ehost-live&scope=site

UN sustainable development goals. (n.d.). Retrieved November 30, 2017, from http://www.un.org/sustainabledevelopment/sustainable-development-goals/

Union, E. (2015). Lund Declaration 2015. Retrieved March 18, 2018, from https://www.vr.se/download/18.43a2830b15168a067b9dac74/1454326776513/The+Lund+Declaration+2015.pdf

van der Ploeg. (2014). The salient history of Dalton education in the Netherlands. *History of Education*, *43*(3), 368–386. https://doi.org/10.1080/0046760X.2014.887792

van de Grift, W. (2007). Quality of teaching in four European countries: a review of the literature and application of an assessment instrument. *Educational Research*, *49*(2), 127–152. https://doi.org/10.1080/00131880701369651

van de Grift, W. J. C. M., Chun, S., Maulana, R., Lee, O., & Helms-Lorenz, M. (2017). Measuring teaching quality and student engagement in South Korea and the Netherlands. *School Effectiveness and School Improvement*, *28*(3), 337–349. https://doi.org/10.1080/09243453.2016.1263215

Vogelzang, J., & Admiraal, W. F. (2017). Classroom action research on formative assessment in a context-based chemistry course. *Educational Action Research*, *25*(1), 155–166. https://doi.org/10.1080/09650792.2016.1177564

Vygotsky, L. S. (1978). *Mind in society: the development of higher psychological processes,*. Cambridge: Harvard University Press.

Wallace, B., Bernardelli, A., Molyneux, C., & Farrell, C. (2012). TASC: thinking actively in a social context. A universal problem-solving process: a powerful tool to promote differentiated learning experiences. *Gifted Education International*, *28*(1), 58–83. https://doi.org/10.1177/0261429411427645

Yayon, M., & Scherz, Z. (2008). The return of the black box. *Journal of Chemical Education*, *85*(4), 541–543. Retrieved from http://search.ebscohost.com.proxy-ub.rug.nl/login.aspx?direct=true&db=eric&AN=EJ826570&site=ehost-live&scope=site

Yusuf, M., Taylor, P. C., & Damanhuri, M. I. M. (2017). Designing critical pedagogy to counteract the hegemonic culture of the traditional chemistry classroom. *Issues in Educational Research*, *27*(1), 168–184. Retrieved from http://search.ebscohost.com.proxy-ub.rug.nl/login.aspx?direct=true&db=eric&AN=EJ1130455&site=ehost-live&scope=site

List of figure sources

https://doi.org/10.1515/9783110569629-012

Index

https://doi.org/10.1515/9783110569629-013